student
ART NOTE

fifth edition

Human Physiology

STUART IRA FOX

Pierce College

WCB **Wm. C. Brown Publishers**

Dubuque, IA Bogota Boston Buenos Aires Caracas Chicago
Guilford, CT London Madrid Mexico City Sydney Toronto

A Times Mirror Company

The credits section for this book begins on page 131 and
is considered an extension of the copyright page.

Copyright © 1996 Times Mirror Higher Education Group, Inc.
All rights reserved

A Times Mirror Company

ISBN 0–697–25383-X

Printed in the United States of America by Wm. C. Brown Communications, Inc.,
2460 Kerper Boulevard, Dubuque, IA 52001

10 9 8 7 6 5 4 3 2

This Student Study Art Notebook is a gratis ancillary to assist students in note taking during lectures. On each page, there are one, two, or sometimes three figures faithfully reproduced from the textbook. Each figure also corresponds to one of the 150 acetates available to instructors who adopt this textbook.

The intention is to place the acetate art in front of students (via the notebook) as the instructor uses the overhead during lectures. The advantage to the student is that he/she will be able to see all labels clearly, and take meaningful notes without having to make hurried sketches of the acetate figure.

The pages of the Art Notebook are perforated and three-hole punched, so they can be removed and placed in a personal binder for specific study and review, or to create space for additional notes.

DIRECTORY OF NOTEBOOK FIGURES

TO ACCOMPANY STUART I. FOX
HUMAN PHYSIOLOGY, 5ED.

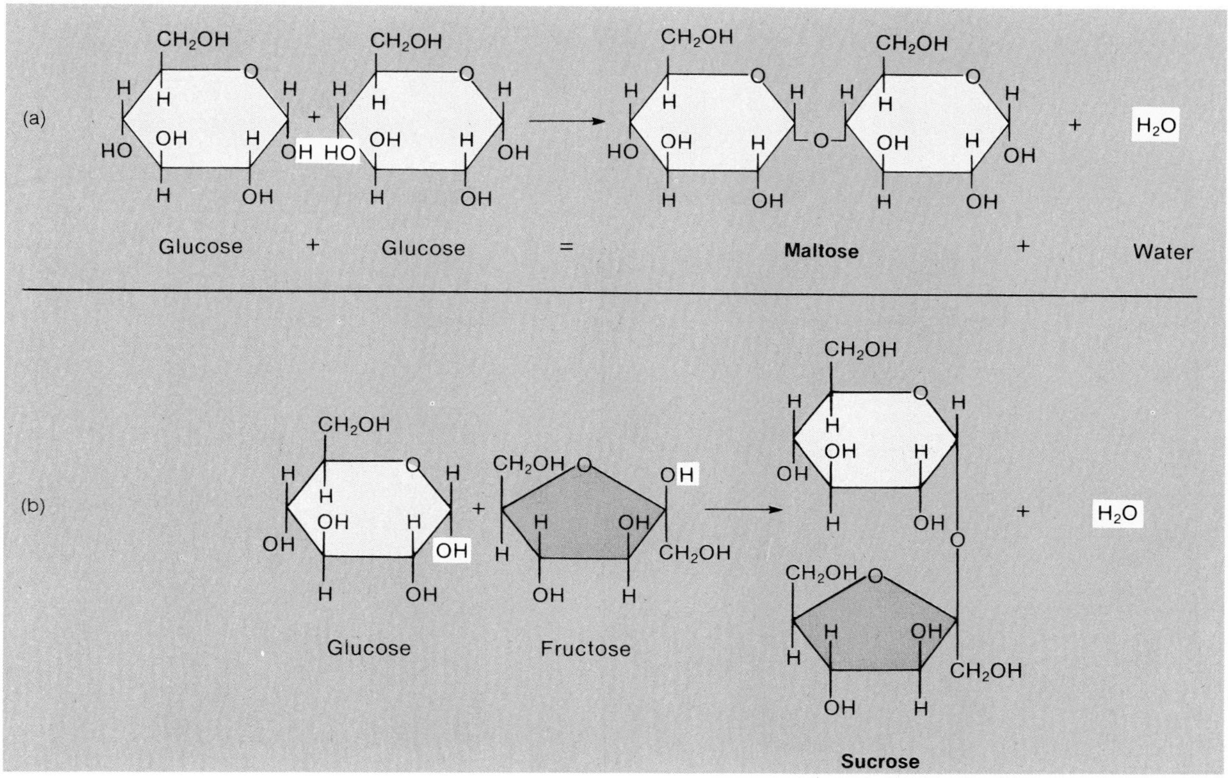

Dehydration Synthesis of Two Disaccharides
Figure 2.15

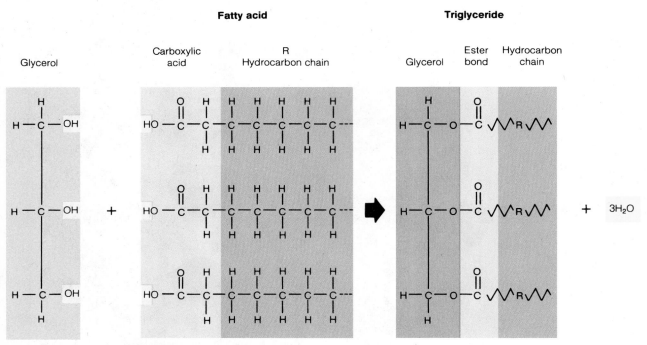

Dehydration Synthesis of a Triglyceride Molecule
Figure 2.18

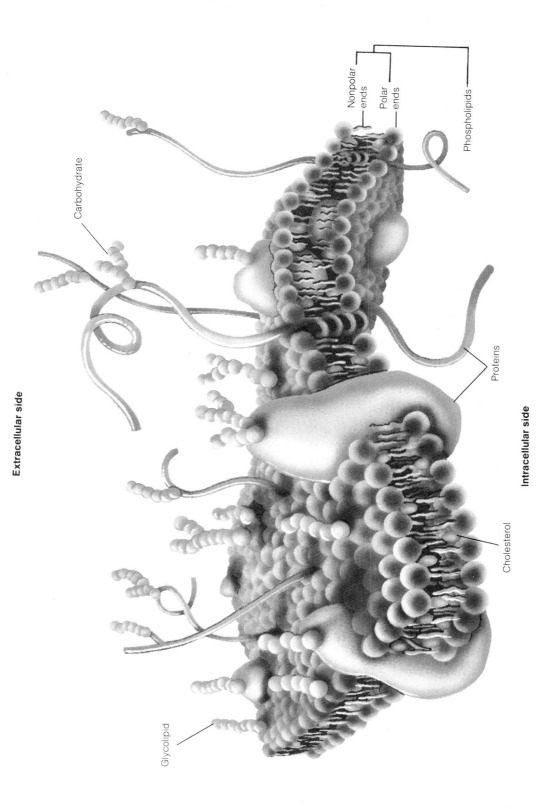

Extracellular side

Intracellular side

Carbohydrate

Glycolipid

Nonpolar ends

Polar ends

Phospholipids

Proteins

Cholesterol

The Fluid-Mosaic Model of the Cell Membrane
Figure 3.2

3

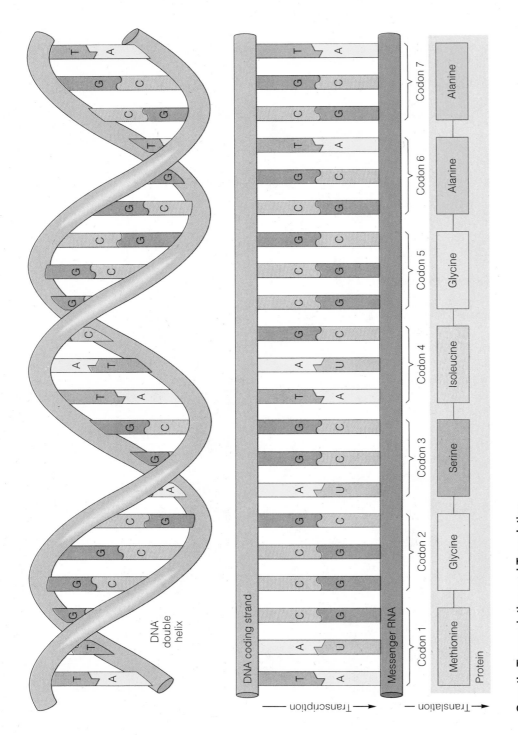

Genetic Transcription and Translation
Figure 3.22

4

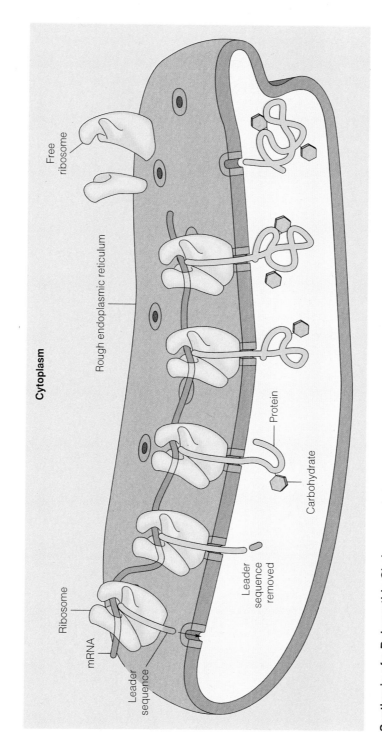

Synthesis of a Polypeptide Chain
Figure 3.25

Cytoplasm

Free ribosome

Rough endoplasmic reticulum

Ribosome

mRNA

Leader sequence

Leader sequence removed

Protein

Carbohydrate

5

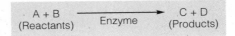

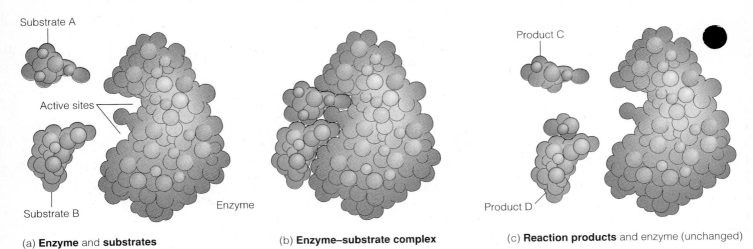

(a) **Enzyme** and **substrates**

(b) **Enzyme–substrate complex**

(c) **Reaction products** and enzyme (unchanged)

The Lock-and-Key Model of Enzyme Action
Figure 4.2

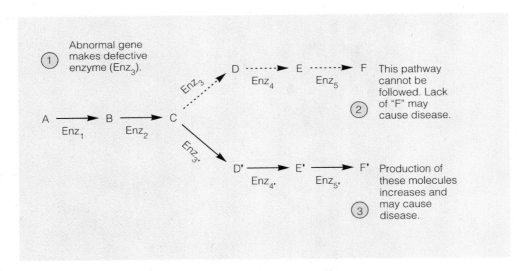

Metabolic Pathways Showing Inborn Errors of Metabolism
Figure 4.10

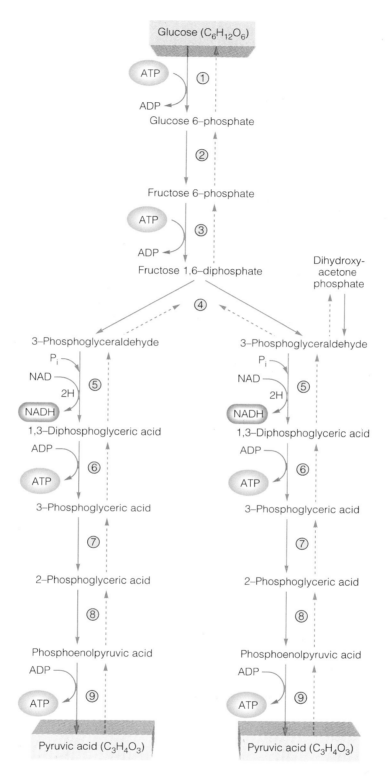
Glucose ($C_6H_{12}O_6$)
ATP
ADP
①
Glucose 6–phosphate
②
Fructose 6–phosphate
ATP
ADP
③
Fructose 1,6–diphosphate
Dihydroxy-
acetone
phosphate
④
3–Phosphoglyceraldehyde
P_i
NAD
2H
NADH
⑤
1,3–Diphosphoglyceric acid
ADP
ATP
⑥
3–Phosphoglyceric acid
⑦
2–Phosphoglyceric acid
⑧
Phosphoenolpyruvic acid
ADP
ATP
⑨
Pyruvic acid ($C_3H_4O_3$)
3–Phosphoglyceraldehyde
P_i
NAD
2H
NADH
⑤
1,3–Diphosphoglyceric acid
ADP
ATP
⑥
3–Phosphoglyceric acid
⑦
2–Phosphoglyceric acid
⑧
Phosphoenolpyruvic acid
ADP
ATP
⑨
Pyruvic acid ($C_3H_4O_3$)

Steps of Glycolysis
Figure 5.2

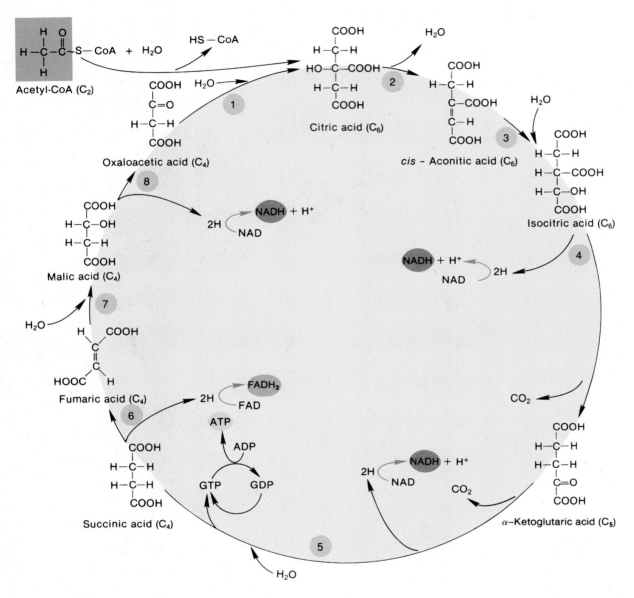
Acetyl-CoA (C_2)
HS — CoA
H_2O
Citric acid (C_6)
cis - Aconitic acid (C_6)
Isocitric acid (C_6)
NADH + H^+
NAD
2H
CO_2
α–Ketoglutaric acid (C_5)
Succinic acid (C_4)
ATP
ADP
GTP
GDP
FADH₂
FAD
Fumaric acid (C_4)
Malic acid (C_4)
Oxaloacetic acid (C_4)
1
2
3
4
5
6
7
8

The Steps of the Krebs Cycle
Figure 5.7

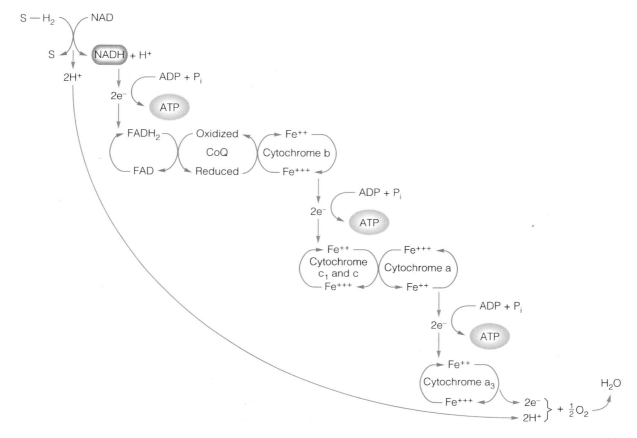

Electron Transport and Oxidative Phosphorylation
Figure 5.8

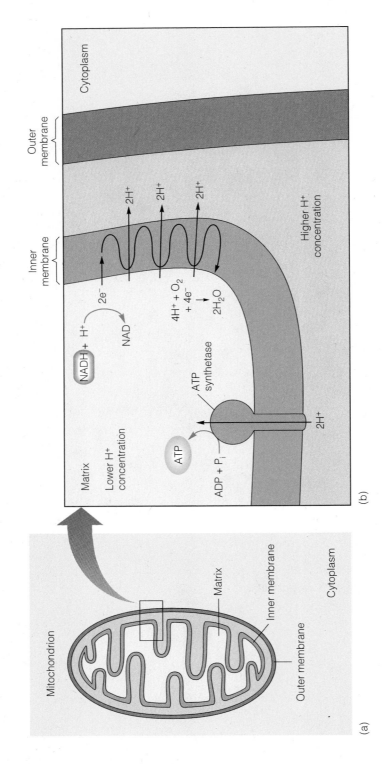

(a)

(b)

The Chemiosmotic Theory of Oxidative Phosphorylation
Figure 5.10

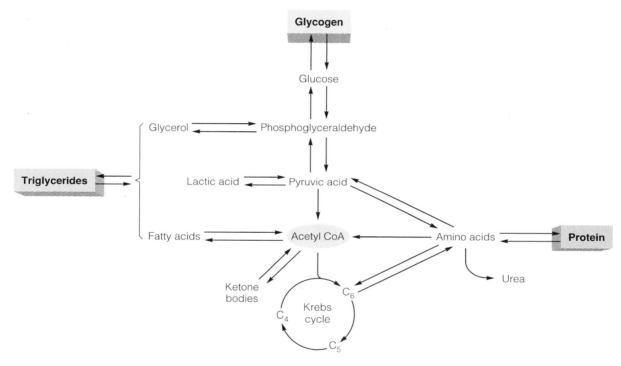

Interrelationships of Glycogen, Fat, and Protein
Figure 5.18

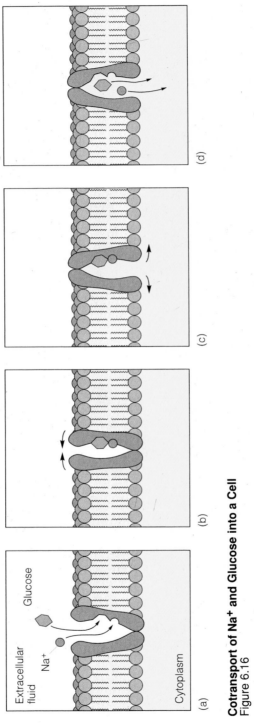

Cotransport of Na⁺ and Glucose into a Cell
Figure 6.16

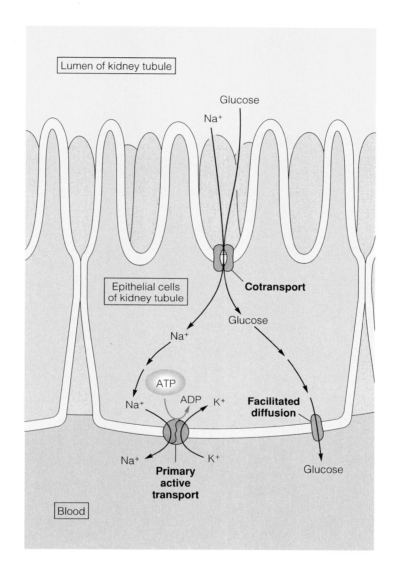

Different Mechanisms of Glucose Transport
Figure 6.17

13

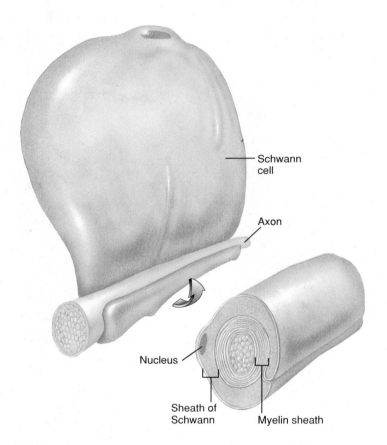

Schwann
cell

Axon

Nucleus

Sheath of
Schwann

Myelin sheath

Formation of Myelin Sheath by Schwann Cells
Figure 7.5

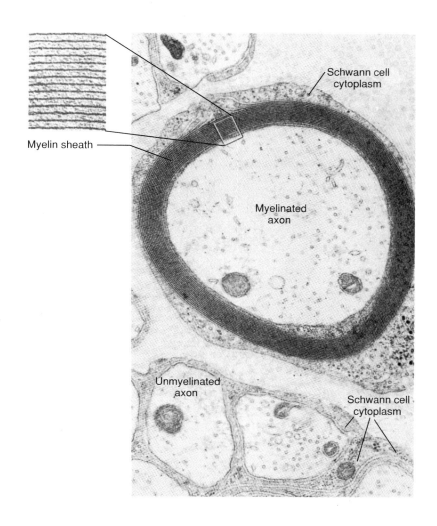

Schwann cell cytoplasm

Myelin sheath

Myelinated axon

Unmyelinated axon

Schwann cell cytoplasm

Unmyelinated and Myelinated Axons
Figure 7.6

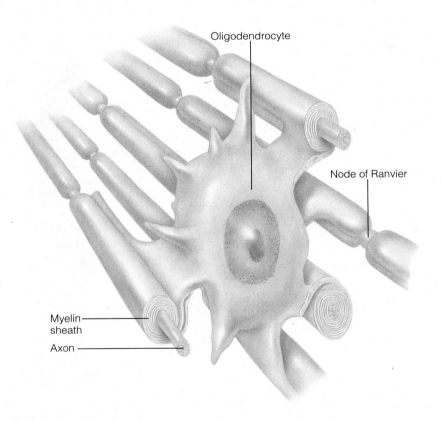

Formation of a Myelin Sheath by Oligodendrocytes
Figure 7.7

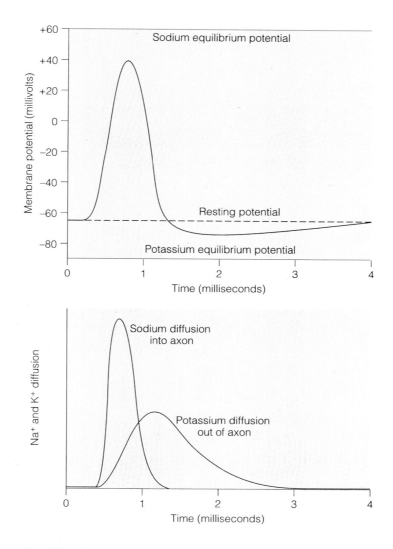

The Diffusion of Na⁺ and K⁺ During an Action Potential
Figure 7.13

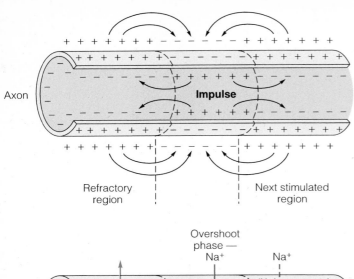

Axon

Impulse

Refractory region

Next stimulated region

Overshoot phase — Na⁺

Na⁺

(Na⁺ gates open)

Axon

K⁺

Previous impulse

Impulse

Next impulse

The Conduction of a Nerve Impulse
Figure 7.16

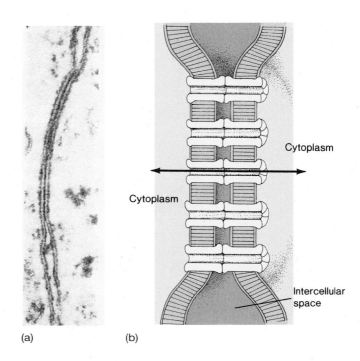

Cytoplasm

Cytoplasm

Intercellular space

(a)   (b)

A Gap Junction
Figure 7.19

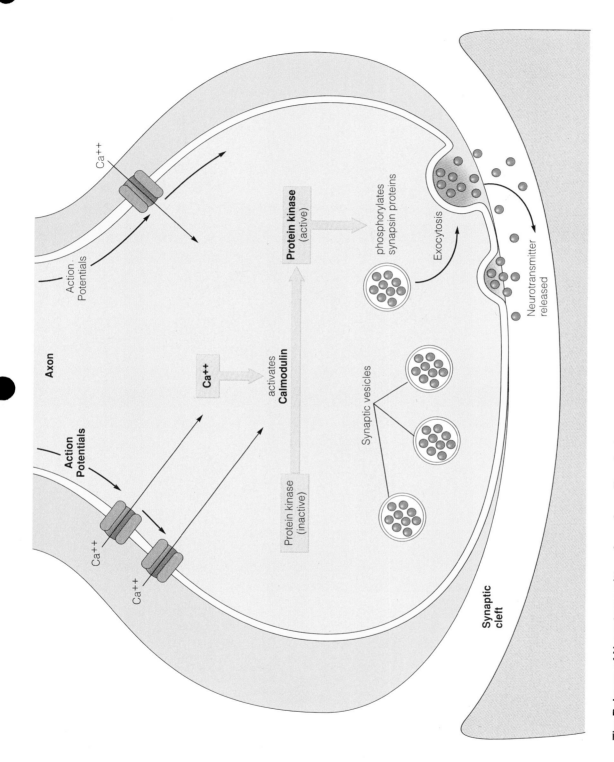

The Release of Neurotransmitters from Axon Terminals
Figure 7.21

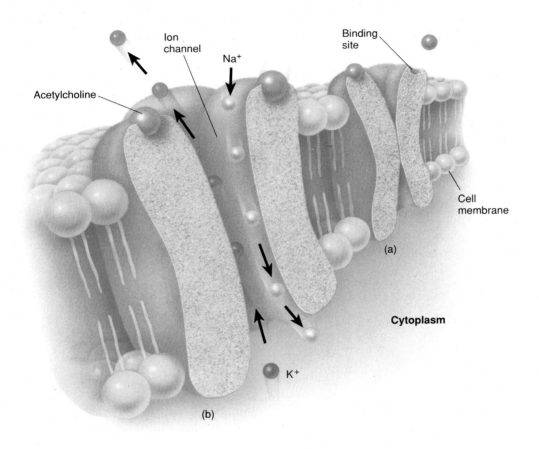

Ion channel

Na⁺

Binding site

Acetylcholine

Cell membrane

(a)

Cytoplasm

(b)

K⁺

Nicotinic Acetylcholine Receptors
Figure 7.22

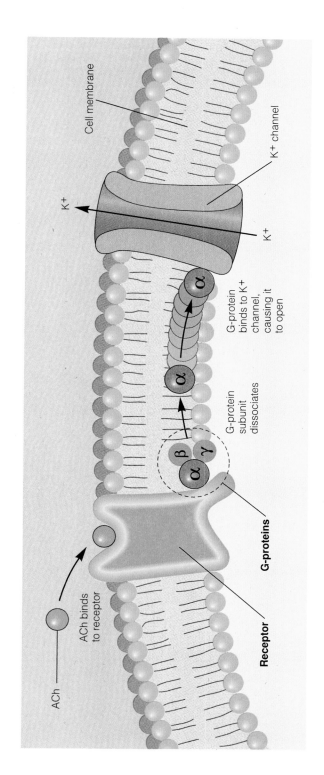

Muscarinic ACh Receptors and G-Proteins
Figure 7.23

21

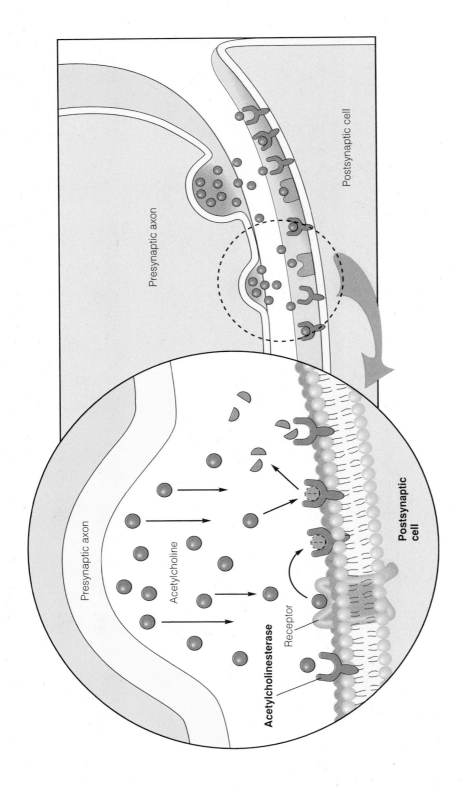

Presynaptic axon

Postsynaptic cell

Presynaptic axon

Acetylcholine

Receptor

Acetylcholinesterase

Postsynaptic cell

AChE in the Postsynaptic Cell Membrane
Figure 7.24

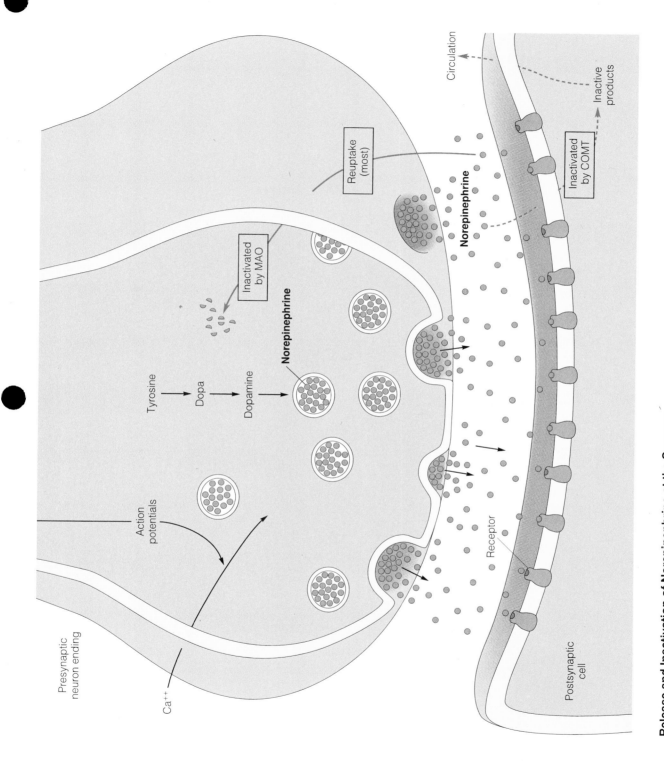

Release and Inactivation of Norepinephrine at the Synapse
Figure 7.27

Circulation

Inactivated
by COMT

Inactive
products

Reuptake
(most)

Norepinephrine

Inactivated
by MAO

Norepinephrine

Tyrosine → Dopa → Dopamine →

Action
potentials

Receptor

Ca++

Presynaptic
neuron ending

Postsynaptic
cell

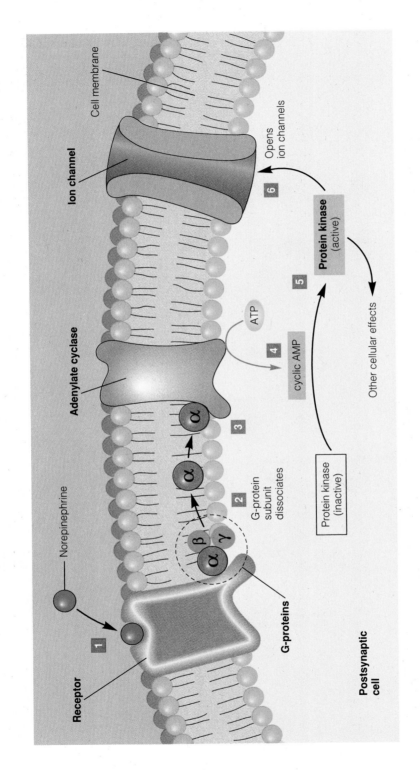

Effect of Norepinephrine in the Postsynaptic Cell
Figure 7.28

24

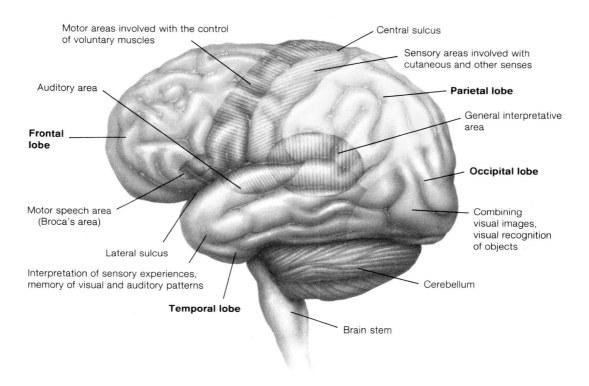

Motor areas involved with the control of voluntary muscles

Central sulcus

Sensory areas involved with cutaneous and other senses

Auditory area

Parietal lobe

Frontal lobe

General interpretative area

Occipital lobe

Motor speech area (Broca's area)

Combining visual images, visual recognition of objects

Lateral sulcus

Interpretation of sensory experiences, memory of visual and auditory patterns

Cerebellum

Temporal lobe

Brain stem

The Lobes of the Left Cerebral Hemisphere
Figure 8.6

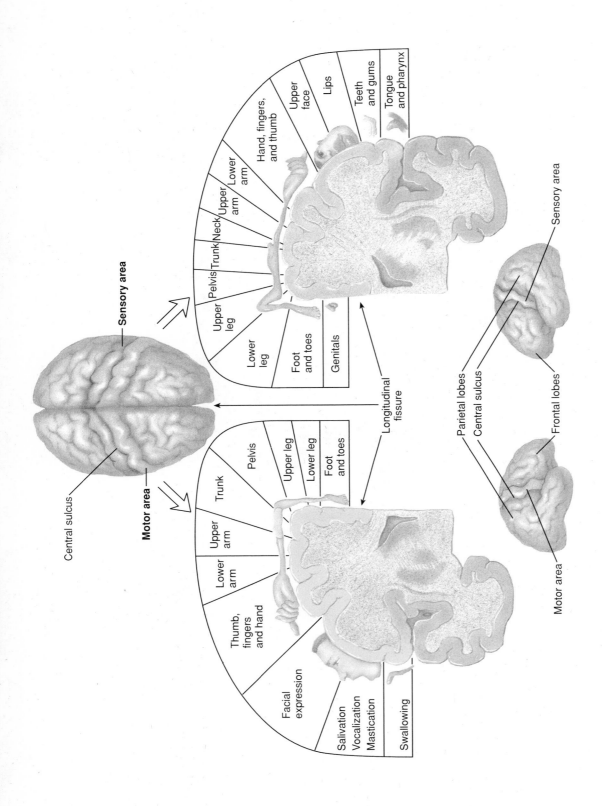

Motor and Sensory Areas of the Cerebral Cortex
Figure 8.7

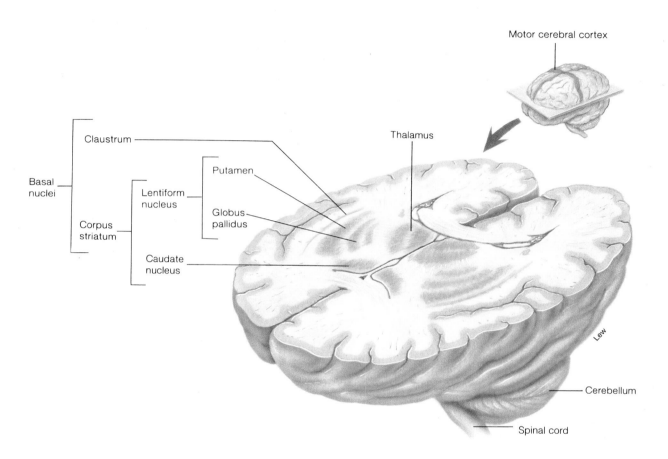

Motor cerebral cortex

Thalamus

Claustrum

Basal nuclei

Putamen

Lentiform nucleus

Corpus striatum

Globus pallidus

Caudate nucleus

Lew

Cerebellum

Spinal cord

The Basal Nuclei
Figure 8.11

27

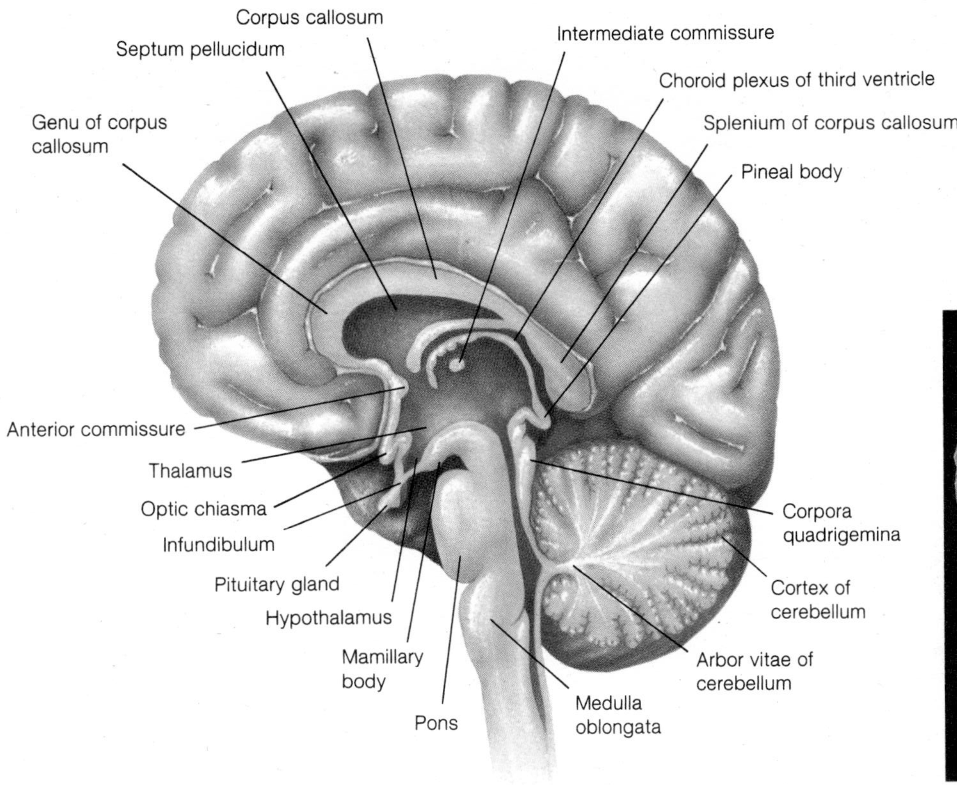

Corpus callosum

Septum pellucidum

Genu of corpus
callosum

Intermediate commissure

Choroid plexus of third ventricle

Splenium of corpus callosum

Pineal body

Corpora
quadrigemina

Cortex of
cerebellum

Arbor vitae of
cerebellum

Medulla
oblongata

Pons

Anterior commissure

Thalamus

Optic chiasma

Infundibulum

Pituitary gland

Hypothalamus

Mamillary
body

(a)

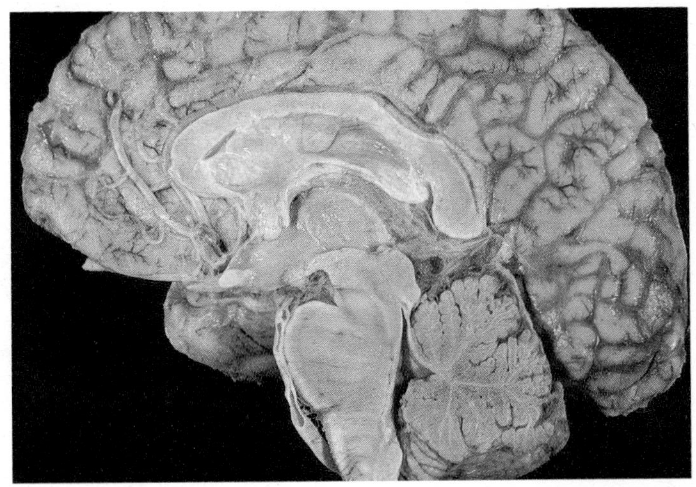

(b)

A Midsagittal Section Through the Brain
Figure 8.15

28

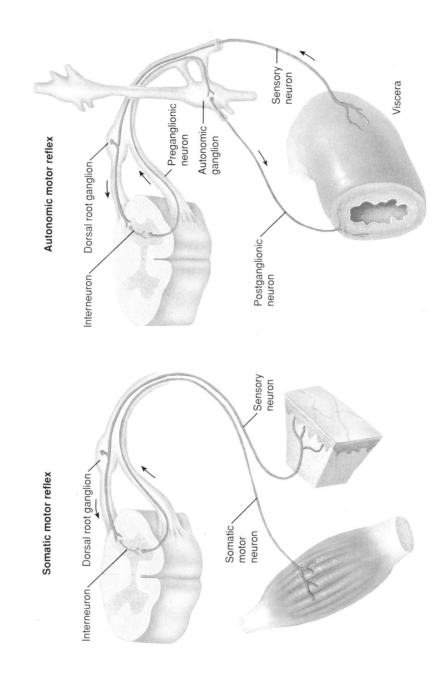

Autonomic motor reflex

Dorsal root ganglion

Interneuron

Preganglionic neuron

Autonomic ganglion

Sensory neuron

Viscera

Postganglionic neuron

Somatic motor reflex

Dorsal root ganglion

Interneuron

Sensory neuron

Somatic motor neuron

Comparison of Somatic Motor and Autonomic Reflex
Figure 9.1

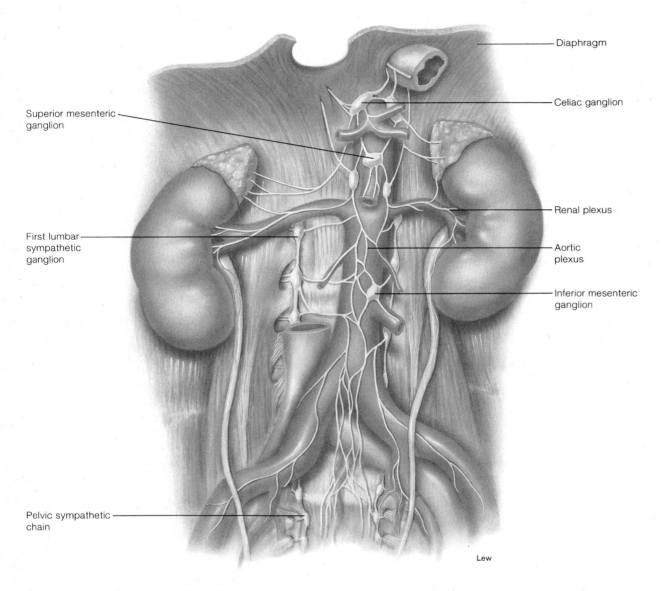

Diaphragm

Celiac ganglion

Superior mesenteric
ganglion

Renal plexus

First lumbar
sympathetic
ganglion

Aortic
plexus

Inferior mesenteric
ganglion

Pelvic sympathetic
chain

Lew

The Collateral Sympathetic Ganglia
Figure 9.5

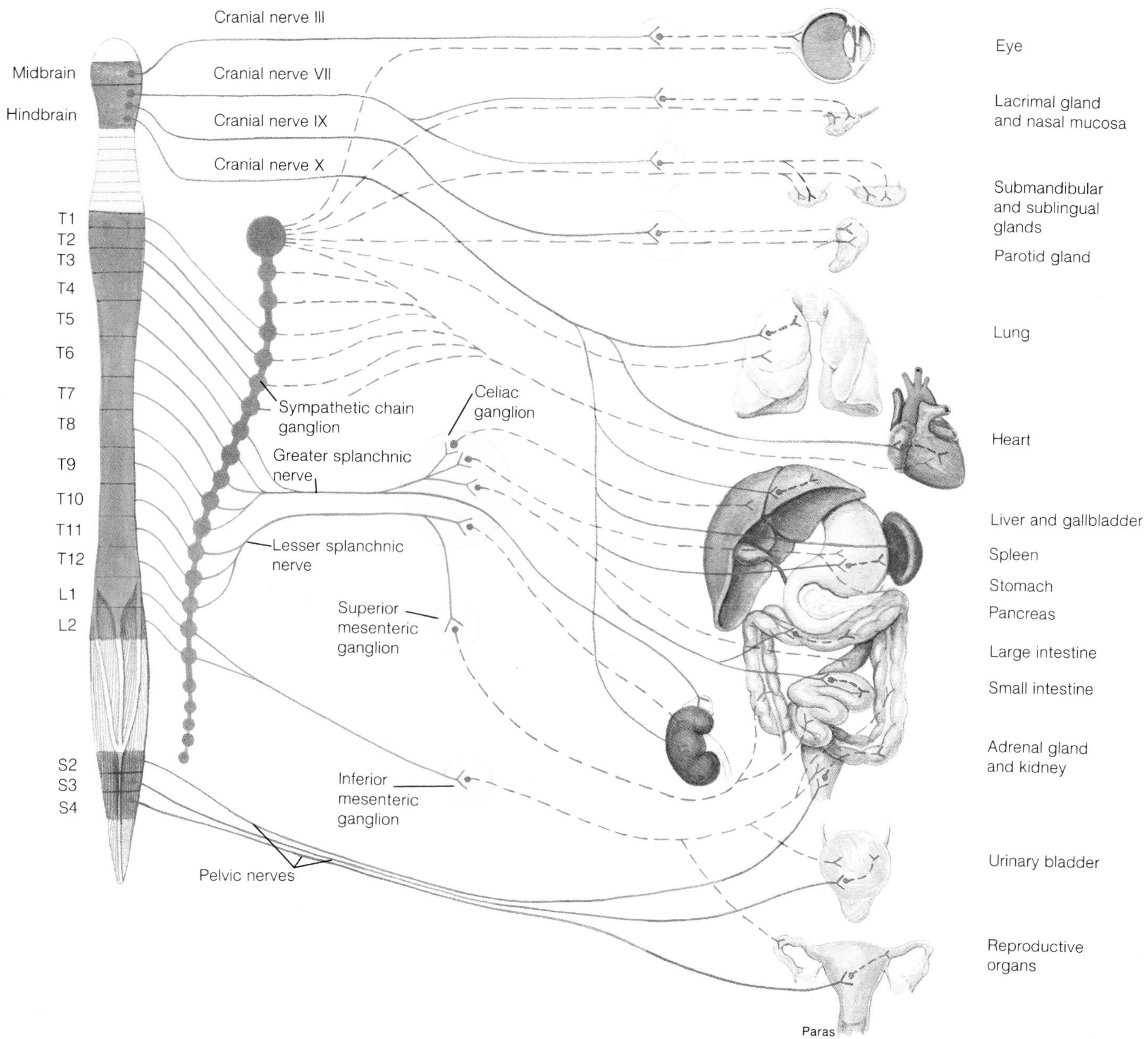

The Autonomic Nervous System
Figure 9.7

Eye

Lacrimal gland and nasal mucosa

Submandibular and sublingual glands

Parotid gland

Lung

Heart

Liver and gallbladder

Spleen

Stomach

Pancreas

Large intestine

Small intestine

Adrenal gland and kidney

Urinary bladder

Reproductive organs

Cranial nerve III

Cranial nerve VII

Cranial nerve IX

Cranial nerve X

Midbrain

Hindbrain

T1
T2
T3
T4
T5
T6
T7
T8
T9
T10
T11
T12
L1
L2

S2
S3
S4

Sympathetic chain ganglion

Greater splanchnic nerve

Lesser splanchnic nerve

Celiac ganglion

Superior mesenteric ganglion

Inferior mesenteric ganglion

Pelvic nerves

Paras

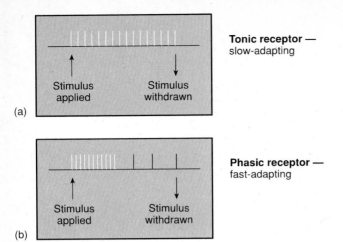

Tonic receptor — slow-adapting

Stimulus applied

Stimulus withdrawn

Phasic receptor — fast-adapting

Stimulus applied

Stimulus withdrawn

(a)

(b)

Tonic and Phasic Receptors
Figure 10.2

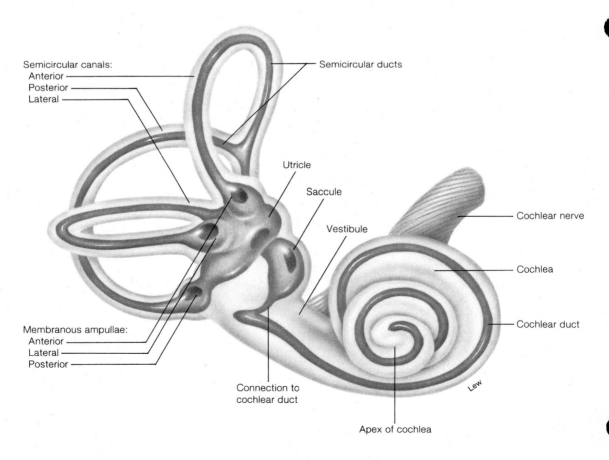

Semicircular canals:
Anterior
Posterior
Lateral

Semicircular ducts

Utricle

Saccule

Vestibule

Cochlear nerve

Cochlea

Cochlear duct

Membranous ampullae:
Anterior
Lateral
Posterior

Connection to cochlear duct

Apex of cochlea

Lew

The Labyrinth of the Inner Ear
Figure 10.13

32

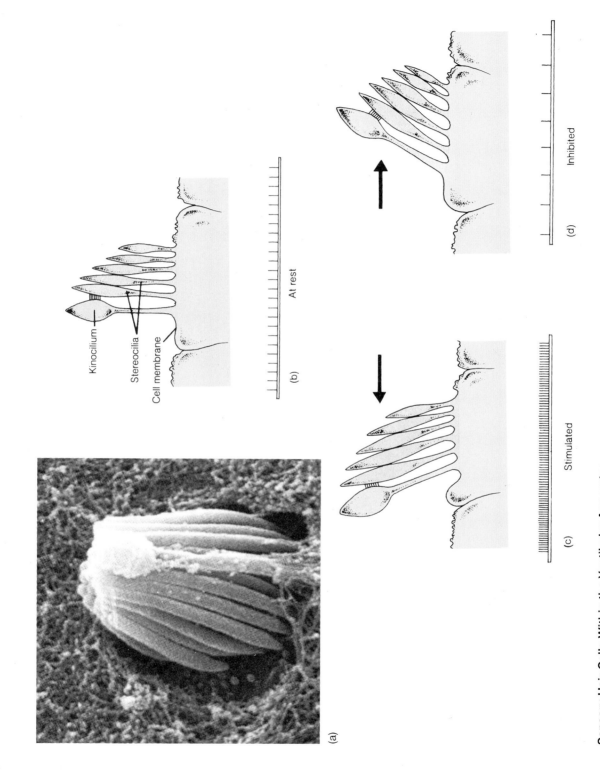

Sensory Hair Cells Within the Vestibular Apparatus
Figure 10.14

Kinocilium

Stereocilia

Cell membrane

(a)

(b) At rest

(c) Stimulated

(d) Inhibited

33

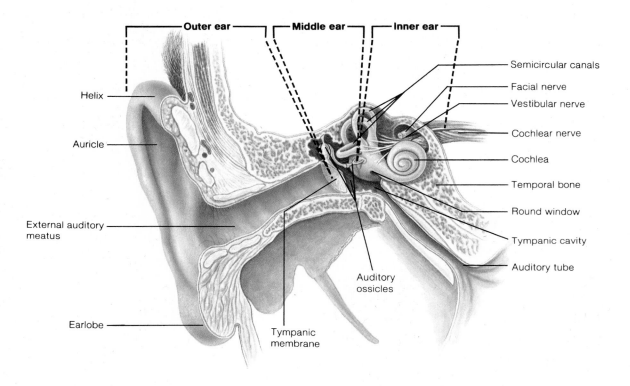

Outer ear — **Middle ear** — **Inner ear**

Helix

Auricle

External auditory
meatus

Earlobe

Tympanic
membrane

Auditory
ossicles

Semicircular canals

Facial nerve

Vestibular nerve

Cochlear nerve

Cochlea

Temporal bone

Round window

Tympanic cavity

Auditory tube

The Ear
Figure 10.18

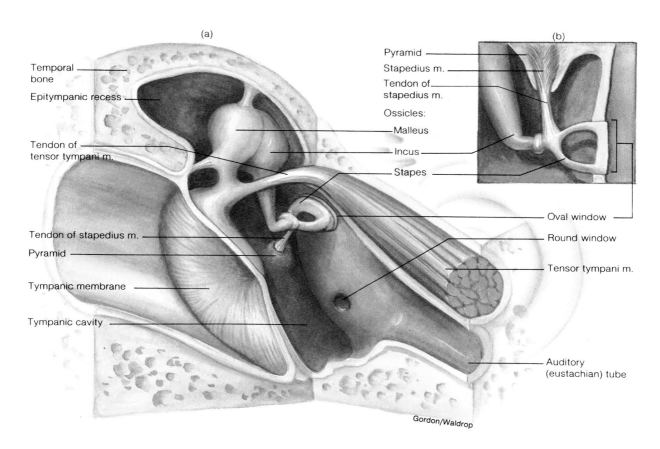

(a)

Temporal bone

Epitympanic recess

Tendon of tensor tympani m.

Tendon of stapedius m.

Pyramid

Tympanic membrane

Tympanic cavity

(b)

Pyramid

Stapedius m.

Tendon of stapedius m.

Ossicles:

Malleus

Incus

Stapes

Oval window

Round window

Tensor tympani m.

Auditory (eustachian) tube

Gordon/Waldrop

The Structures of the Middle Ear
Figure 10.19

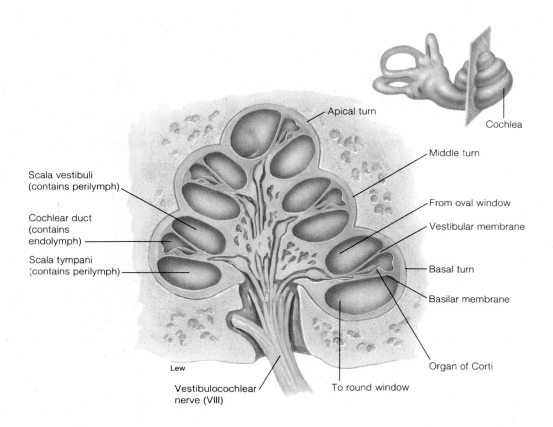

Apical turn

Cochlea

Middle turn

Scala vestibuli
(contains perilymph)

From oval window

Vestibular membrane

Cochlear duct
(contains
endolymph)

Scala tympani
(contains perilymph)

Basal turn

Basilar membrane

Lew

Organ of Corti

Vestibulocochlear
nerve (VIII)

To round window

The Structure of the Cochlea
Figure 10.20

36

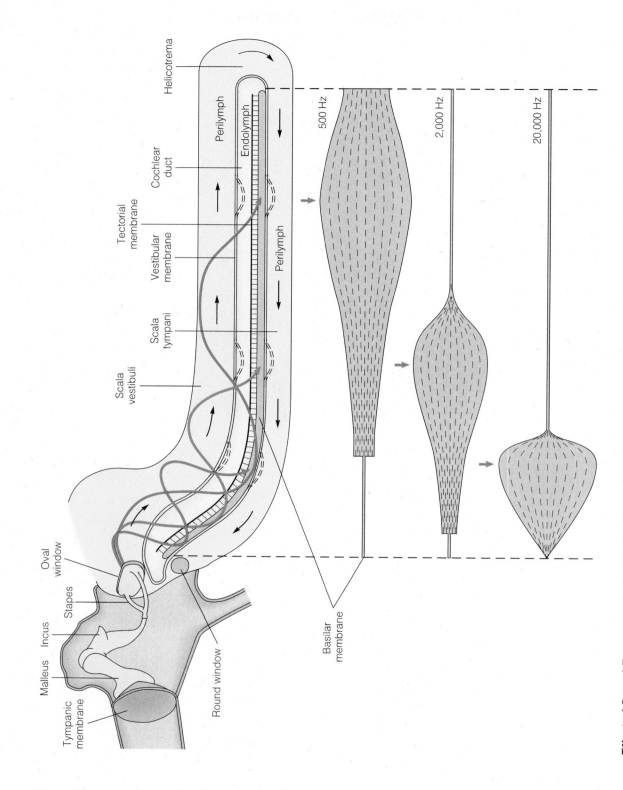

Effect of Sound Frequency on the Basilar Membrane
Figure 10.21

Helicotrema

Perilymph

Cochlear
duct

Endolymph

Tectorial
membrane

500 Hz

Vestibular
membrane

Perilymph

2,000 Hz

Scala
tympani

20,000 Hz

Scala
vestibuli

Oval
window

Stapes

Incus

Basilar
membrane

Malleus

Round window

Tympanic
membrane

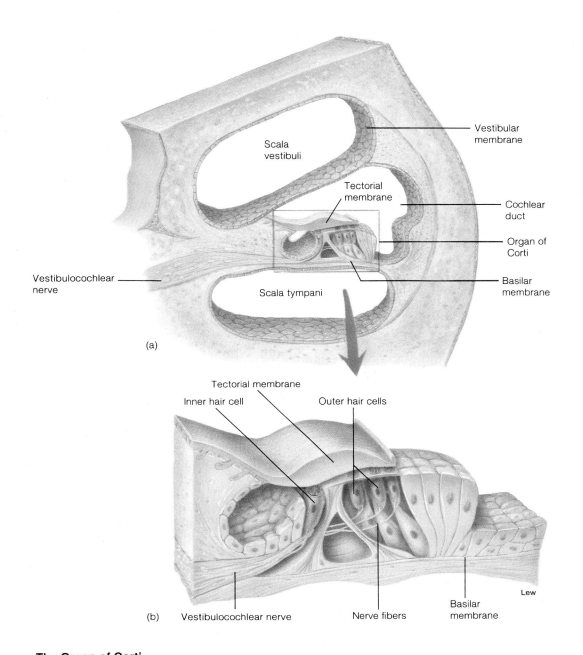

Scala vestibuli

Vestibular membrane

Tectorial membrane

Cochlear duct

Organ of Corti

Basilar membrane

Vestibulocochlear nerve

Scala tympani

(a)

Tectorial membrane

Inner hair cell

Outer hair cells

Vestibulocochlear nerve

Nerve fibers

Basilar membrane

Lew

(b)

The Organ of Corti
Figure 10.23

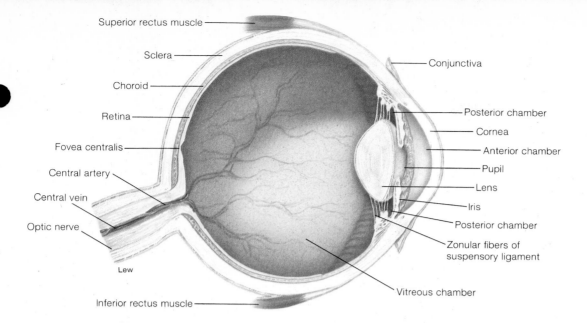

The Internal Anatomy of the Eyeball
Figure 10.27

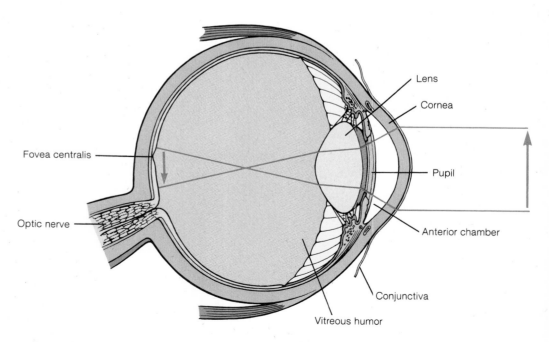

Refraction of Light by Structures of the Eye
Figure 10.31

39

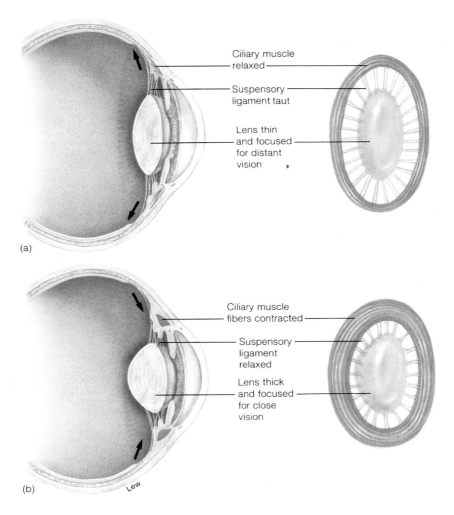

Ciliary muscle
relaxed

Suspensory
ligament taut

Lens thin
and focused
for distant
vision

(a)

Ciliary muscle
fibers contracted

Suspensory
ligament
relaxed

Lens thick
and focused
for close
vision

(b)

Changes in the Shape of the Lens During Accommodation
Figure 10.34

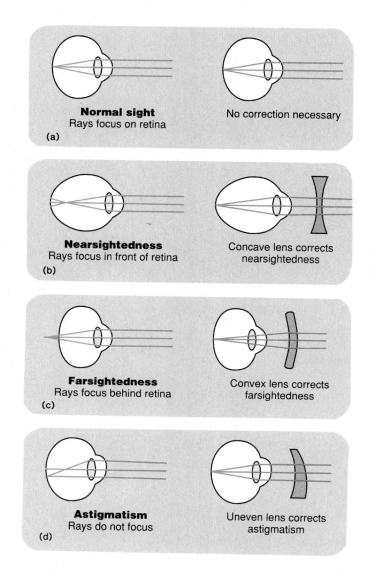

Normal sight
Rays focus on retina
(a)

No correction necessary

Nearsightedness
Rays focus in front of retina
(b)

Concave lens corrects
nearsightedness

Farsightedness
Rays focus behind retina
(c)

Convex lens corrects
farsightedness

Astigmatism
Rays do not focus
(d)

Uneven lens corrects
astigmatism

The Ability of the Eyes to Focus Light
Figure 10.35

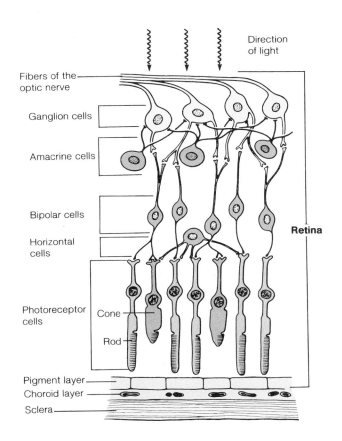

Direction of light

Fibers of the optic nerve

Ganglion cells

Amacrine cells

Bipolar cells

Horizontal cells

Retina

Photoreceptor cells

Cone

Rod

Pigment layer

Choroid layer

Sclera

The Layers of the Retina
Figure 10.36

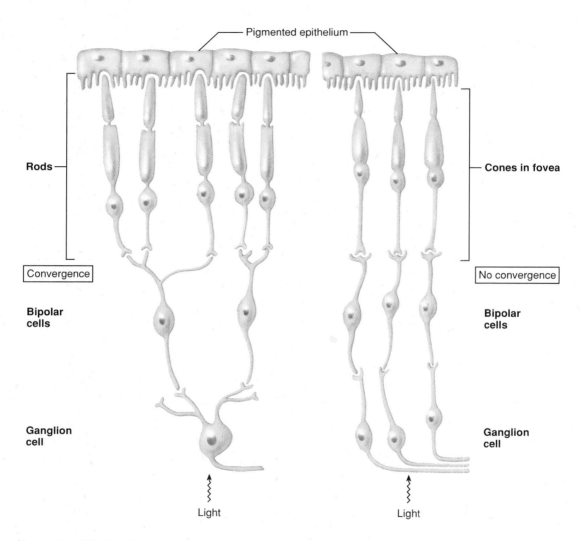

Pigmented epithelium

Rods

Cones in fovea

Convergence

No convergence

Bipolar cells

Bipolar cells

Ganglion cell

Ganglion cell

Light

Light

Cones Provide Acuity, Rods Provide Sensitivity
Figure 10.41

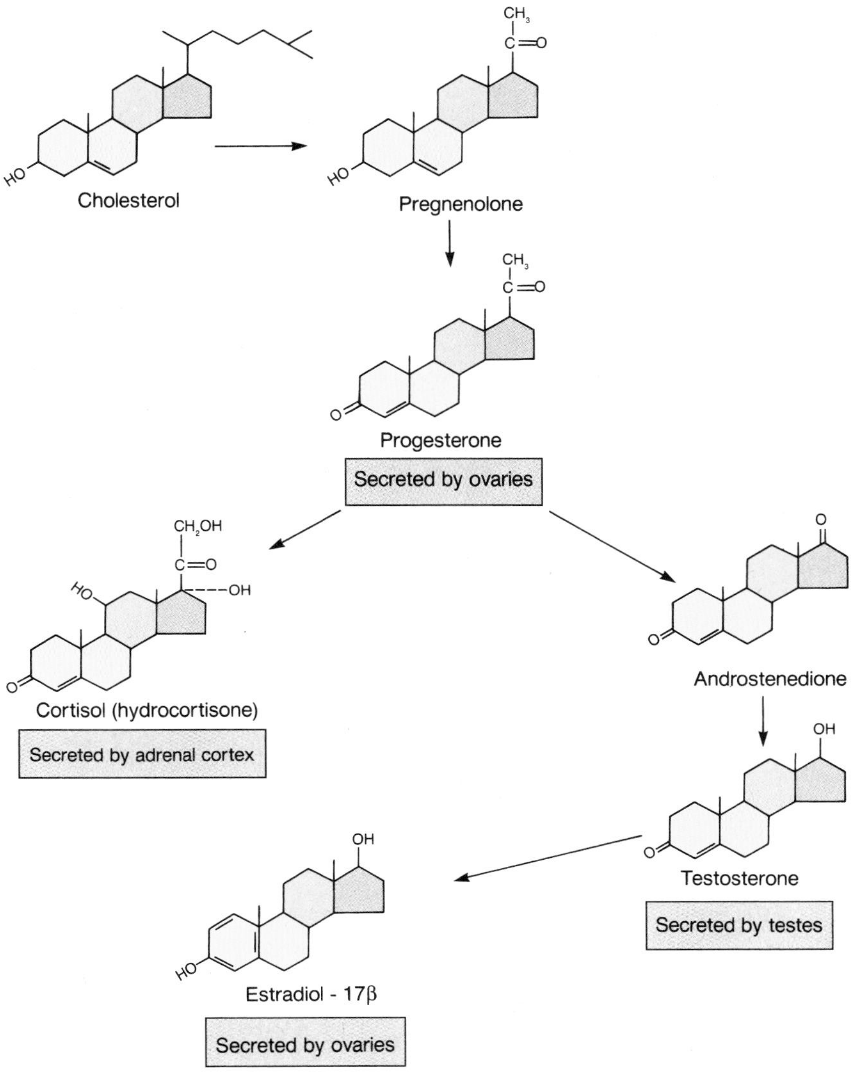

Simplified Biosynthetic Pathways for Steroid Hormones
Figure 11.2

Cholesterol

Pregnenolone

Progesterone

Secreted by ovaries

Cortisol (hydrocortisone)

Secreted by adrenal cortex

Androstenedione

Testosterone

Secreted by testes

Estradiol - 17β

Secreted by ovaries

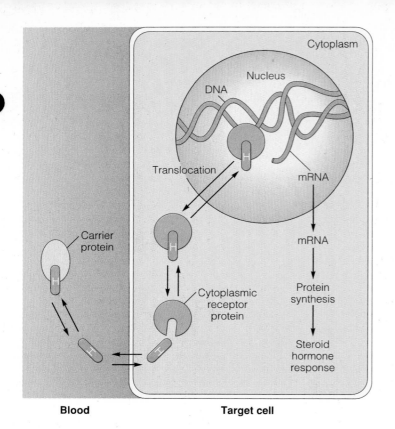

Mechanism of Action of Steroid Hormones
Figure 11.4

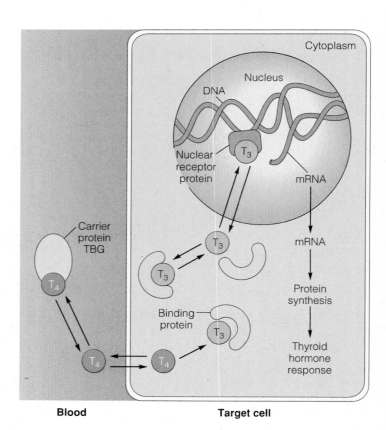

The Mechanism of the Action of T3 on the Target Cells
Figure 11.5

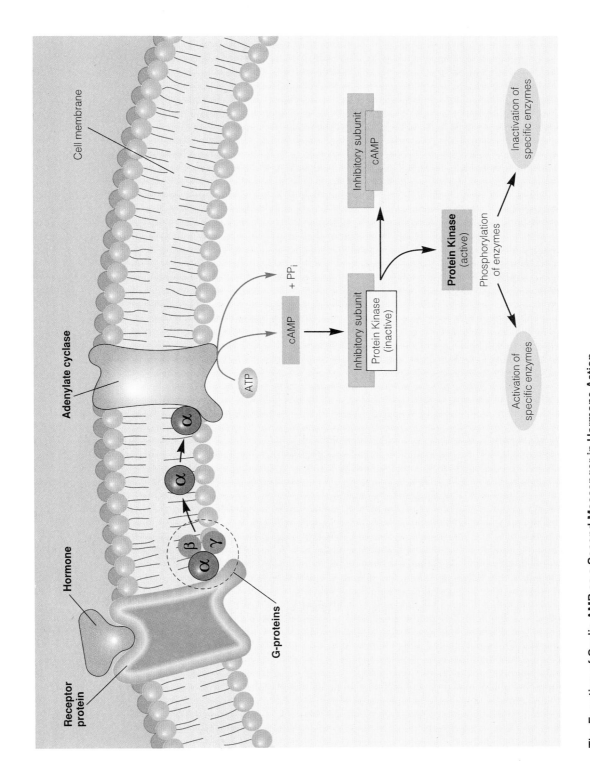

Cell membrane

Adenylate cyclase

Hormone

Receptor protein

G-proteins

α

α

β γ

α

ATP

+ PP$_i$

cAMP

cAMP

Inhibitory subunit

Inhibitory subunit
Protein Kinase
(inactive)

Inhibitory subunit
cAMP

Protein Kinase
(active)

Phosphorylation
of enzymes

Inactivation of
specific enzymes

Activation of
specific enzymes

The Function of Cyclic AMP as a Second Messenger in Hormone Action
Figure 11.6

46

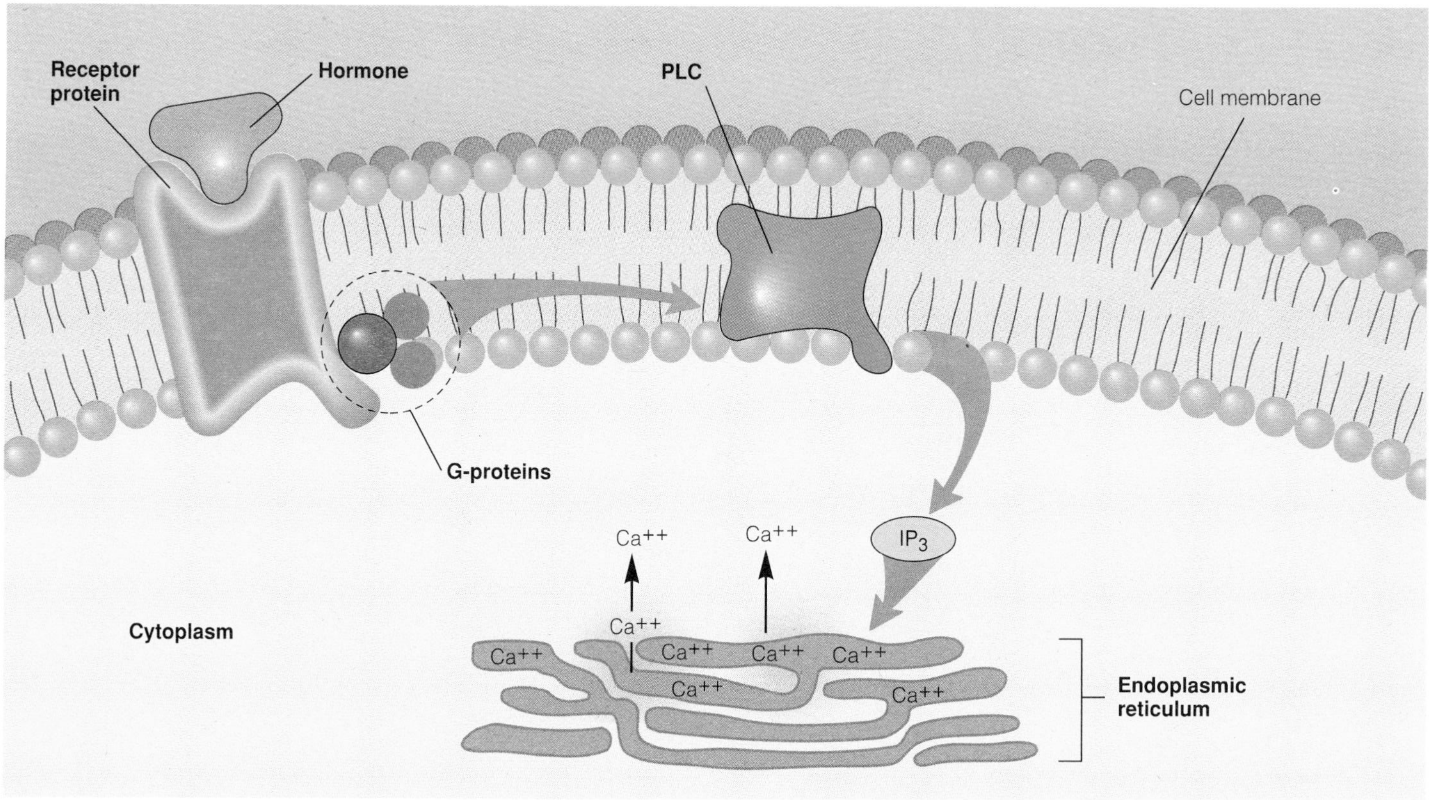

Hormones Activating PLC Use Ca++ as a Second Messanger
Figure 11.7

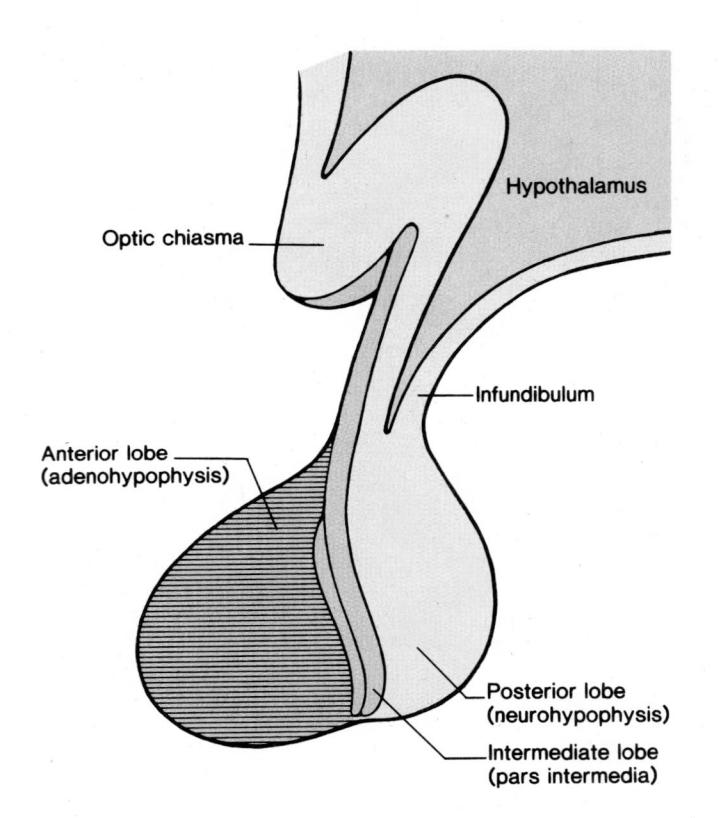
Hypothalamus
Optic chiasma
Infundibulum
Anterior lobe
(adenohypophysis)
Posterior lobe
(neurohypophysis)
Intermediate lobe
(pars intermedia)

Structure of Pituitary Gland
Figure 11.9

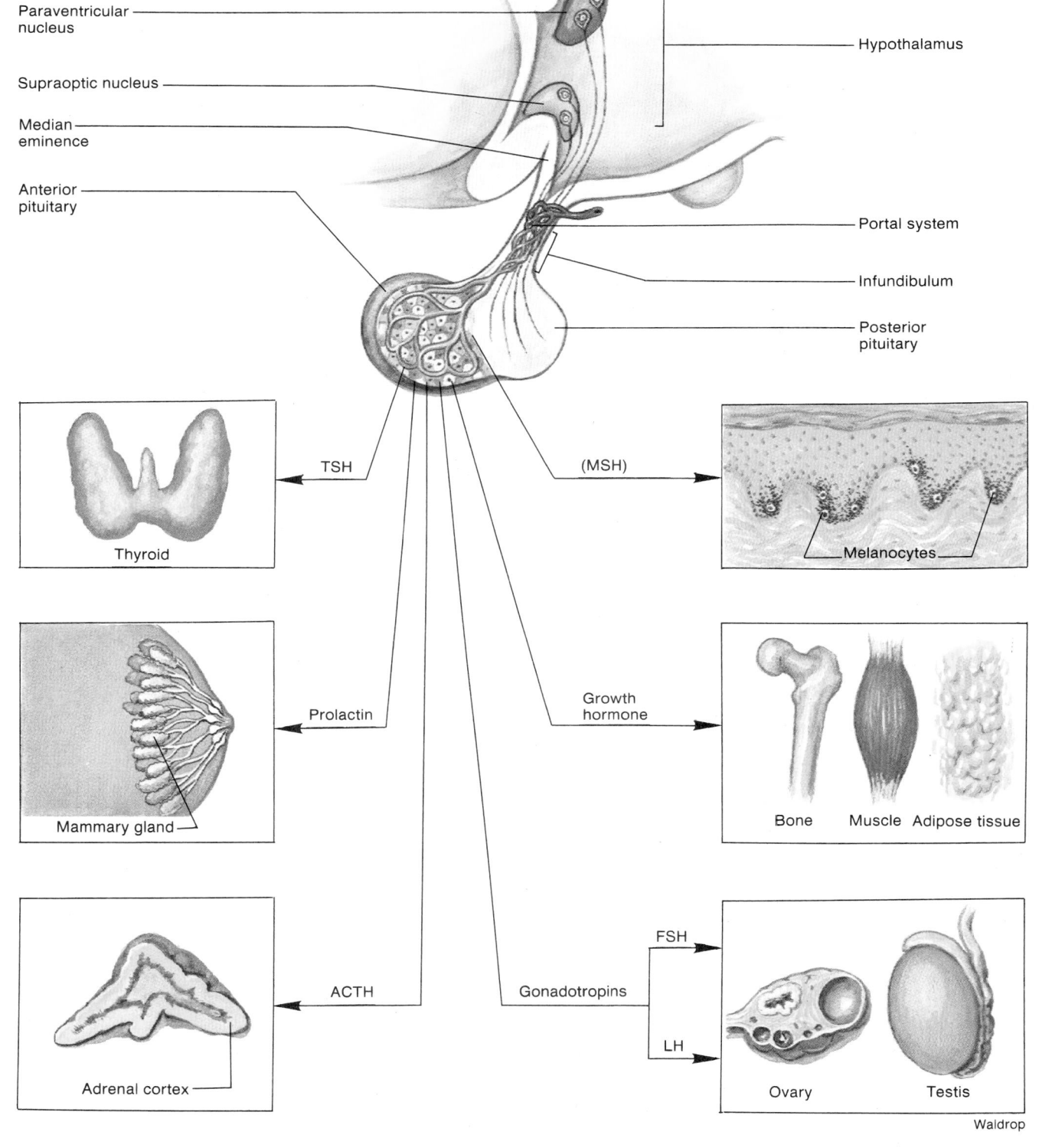

Paraventricular nucleus

Supraoptic nucleus

Median eminence

Anterior pituitary

Hypothalamus

Portal system

Infundibulum

Posterior pituitary

TSH

Thyroid

(MSH)

Melanocytes

Prolactin

Mammary gland

Growth hormone

Bone　Muscle　Adipose tissue

ACTH

Adrenal cortex

Gonadotropins

FSH

LH

Ovary　Testis

Waldrop

Hormones Secreted by the Anterior Pituitary
Figure 11.11

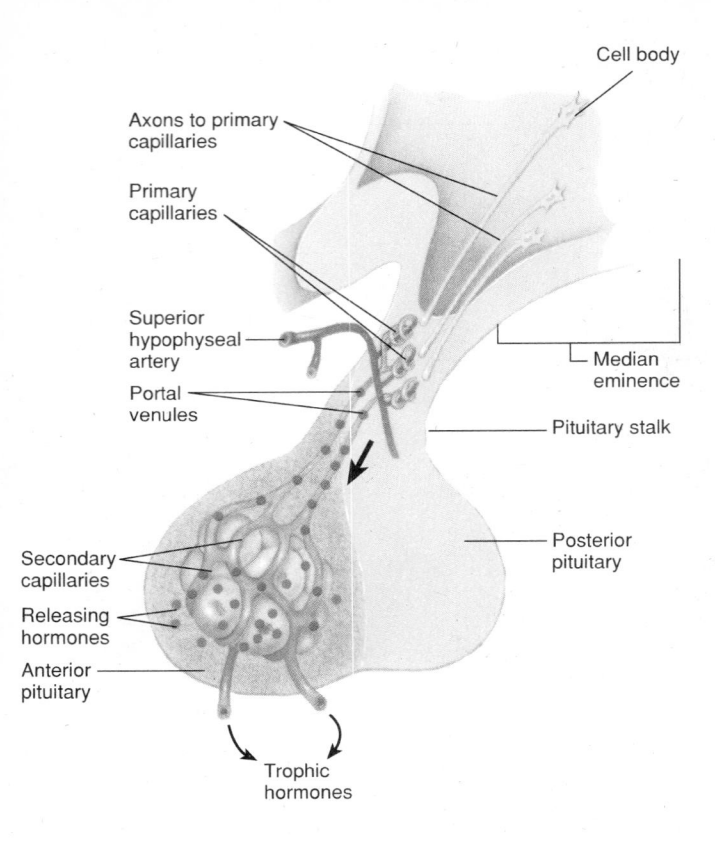

Cell body

Axons to primary
capillaries

Primary
capillaries

Superior
hypophyseal
artery

Portal
venules

Median
eminence

Pituitary stalk

Posterior
pituitary

Secondary
capillaries

Releasing
hormones

Anterior
pituitary

Trophic
hormones

**Control of the Anterior Pituitary by Hormones Secreted by the
Hypothalamus into Portal Blood Vessels**
Figure 11.12

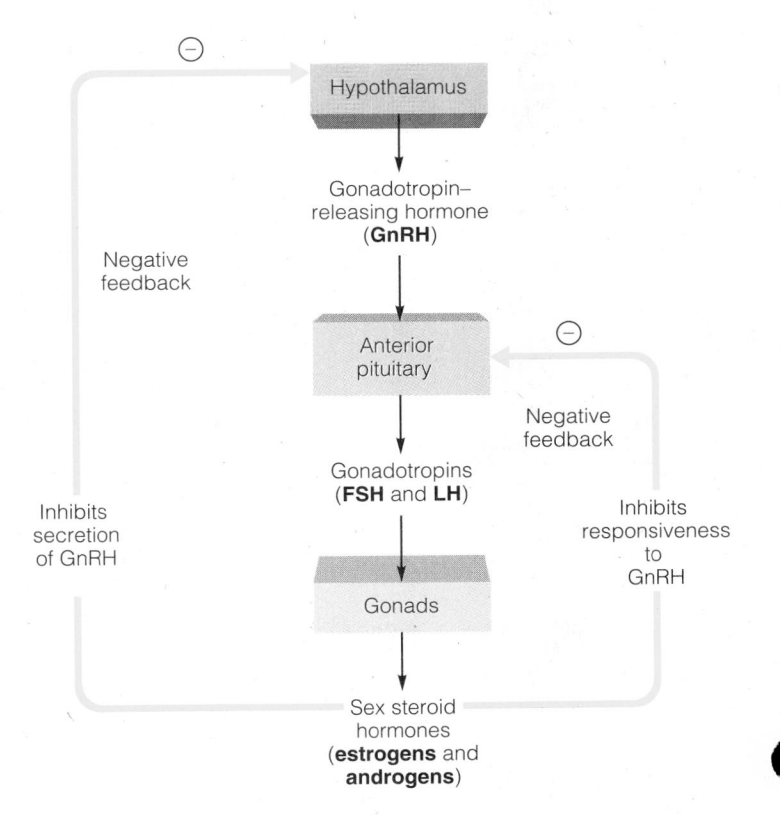

Hypothalamus

Gonadotropin–
releasing hormone
(**GnRH**)

Negative
feedback

Anterior
pituitary

Negative
feedback

Gonadotropins
(**FSH** and **LH**)

Inhibits
secretion
of GnRH

Inhibits
responsiveness
to
GnRH

Gonads

Sex steroid
hormones
(**estrogens** and
androgens)

Negative Feedback Control of Gonadotropin Secretion
Figure 11.14

50

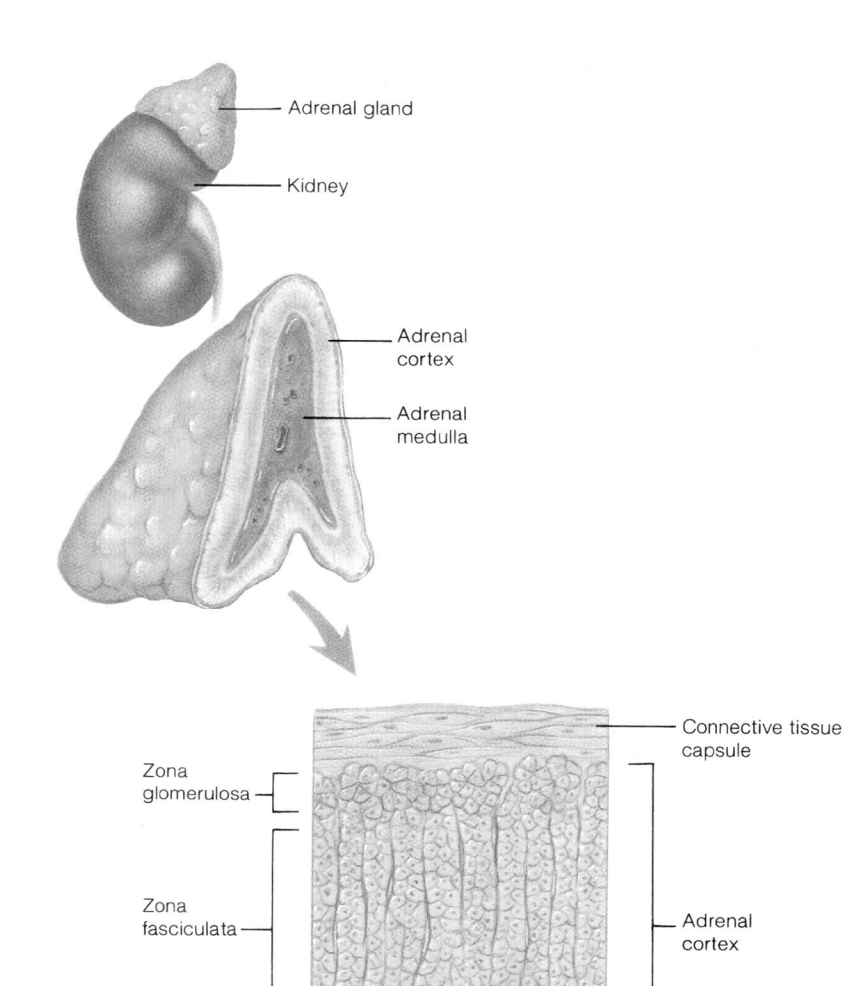
Adrenal gland
Kidney
Adrenal cortex
Adrenal medulla
Connective tissue capsule
Zona glomerulosa
Zona fasciculata
Zona reticularis
Adrenal cortex
Adrenal medulla
Lew

Structure of Adrenal Gland
Figure 11.15

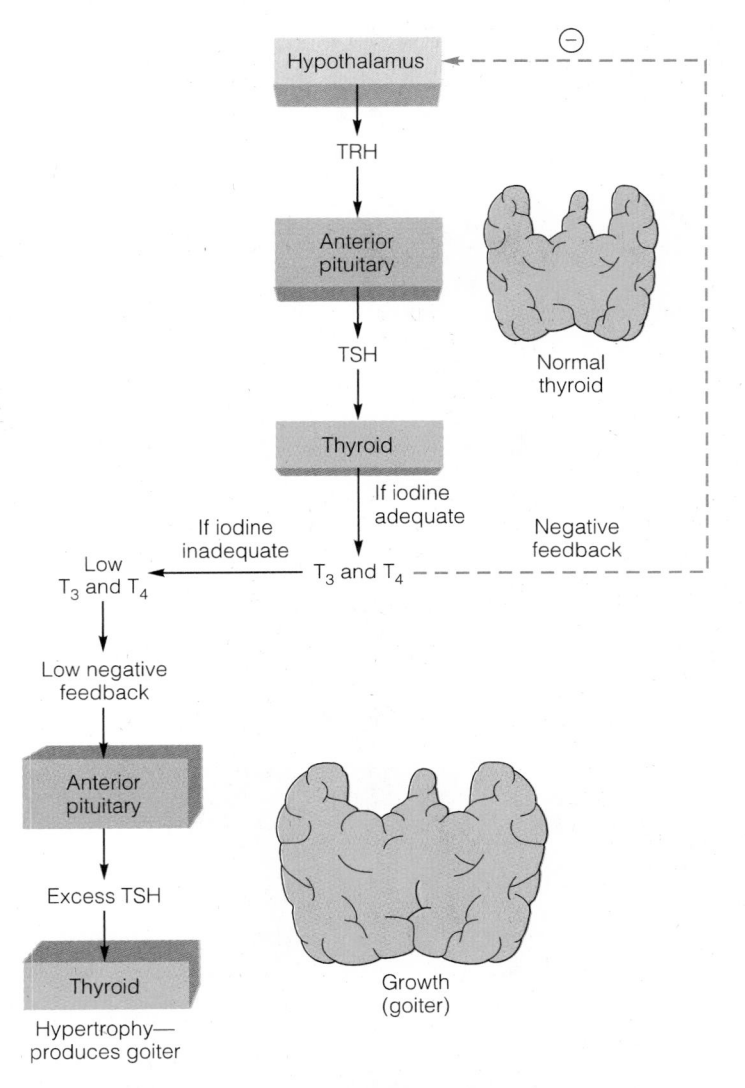

Lack of Adequate Iodine in the Diet Interferes with the Negative Feedback Control of the TSH Secretion
Figure 11.22

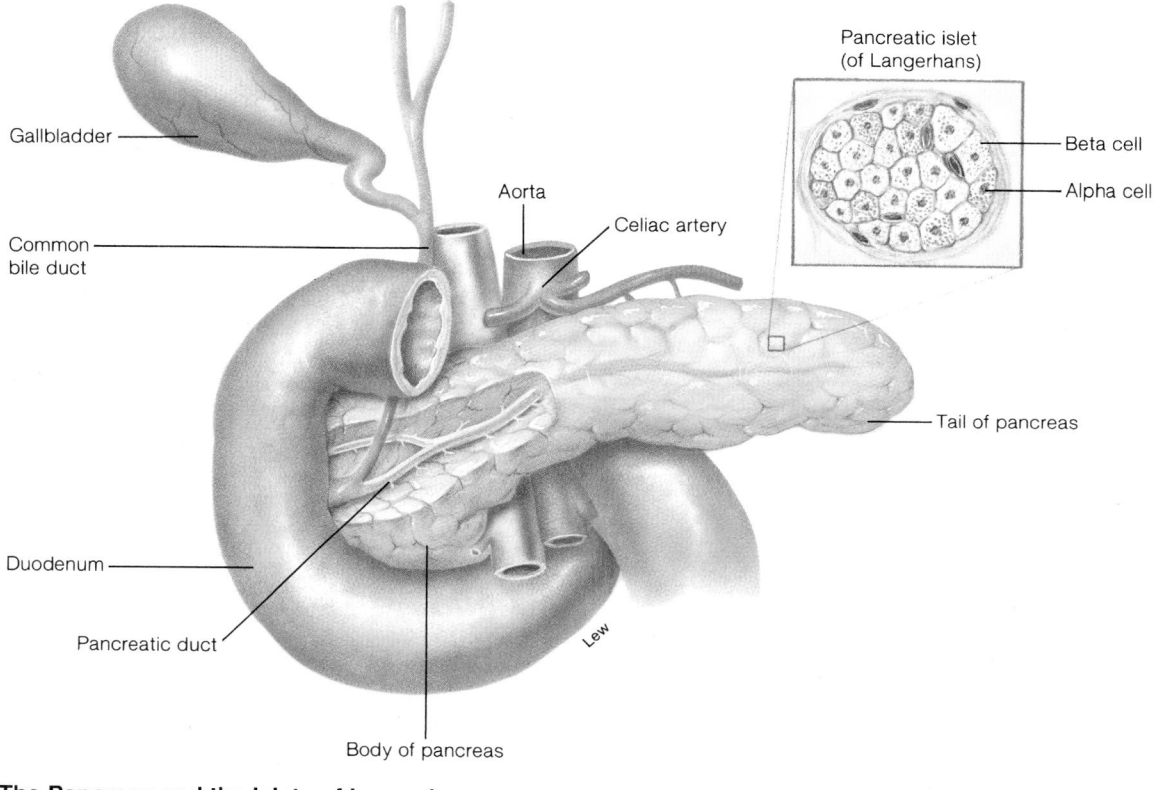

Gallbladder

Aorta

Celiac artery

Common bile duct

Pancreatic islet (of Langerhans)

Beta cell

Alpha cell

Tail of pancreas

Duodenum

Pancreatic duct

Lew

Body of pancreas

The Pancreas and the Islets of Langerhans
Figure 11.26

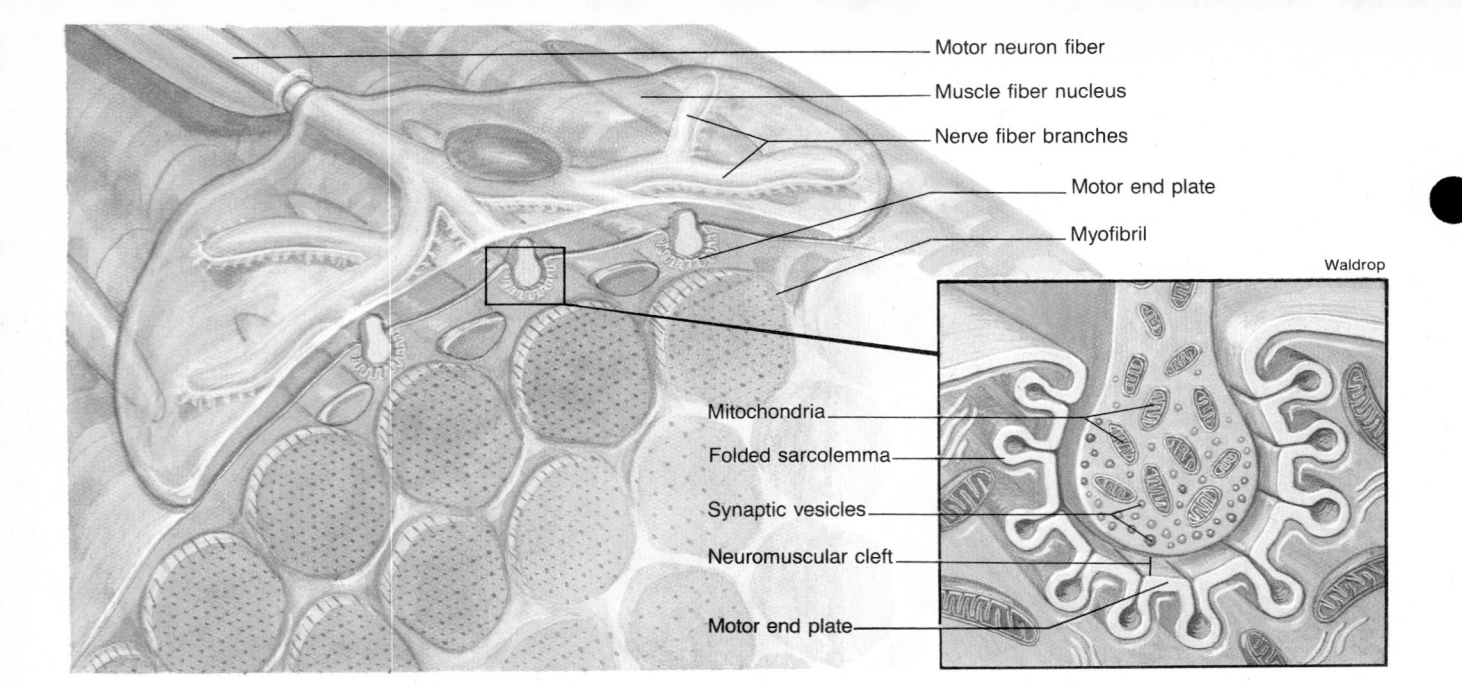

Motor neuron fiber
Muscle fiber nucleus
Nerve fiber branches
Motor end plate
Myofibril

Waldrop

Mitochondria
Folded sarcolemma
Synaptic vesicles
Neuromuscular cleft
Motor end plate

(a)

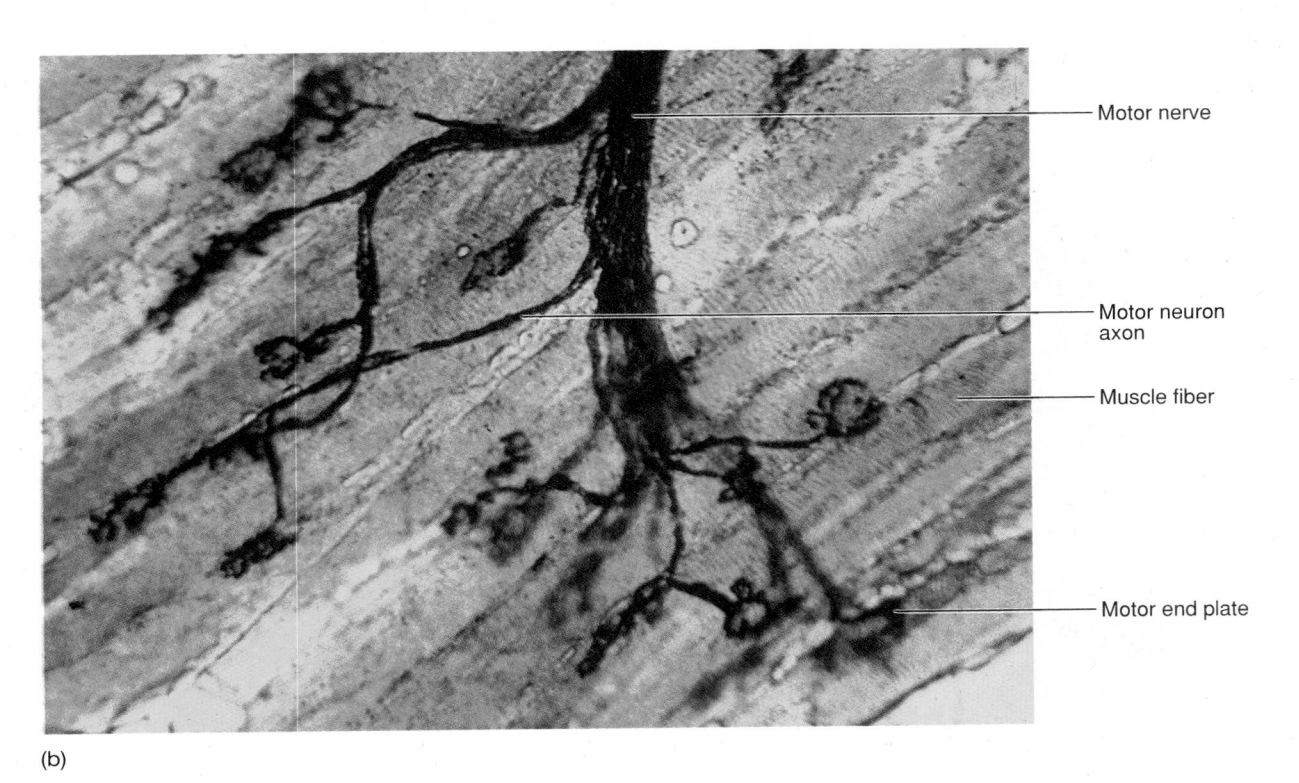

Motor nerve

Motor neuron axon

Muscle fiber

Motor end plate

(b)

The Motor End Plate
Figure 12.6

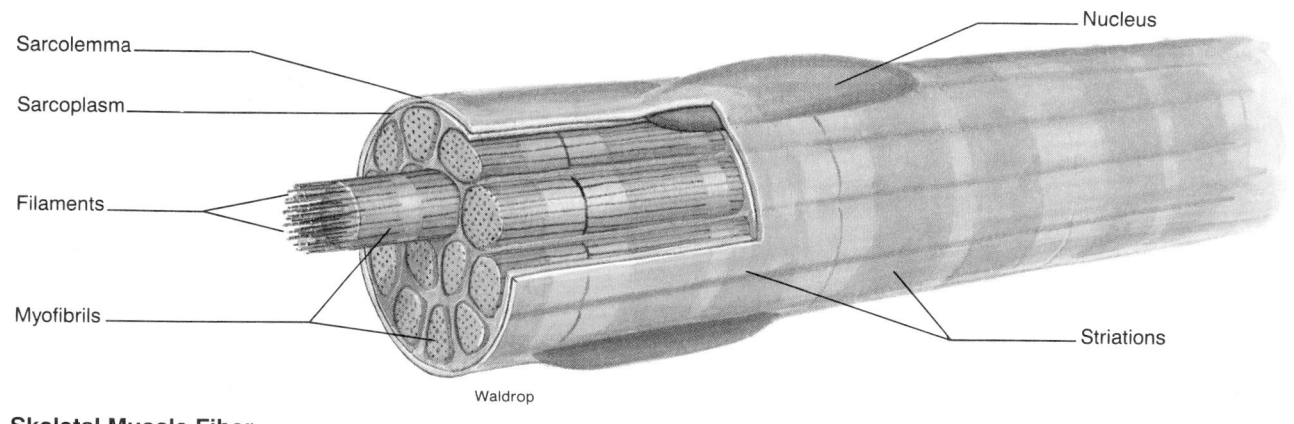

Sarcolemma

Sarcoplasm

Filaments

Myofibrils

Nucleus

Striations

Waldrop

Skeletal Muscle Fiber
Figure 12.8

Myofibril

Myofibril

Myofibrils Showing A, H, and I Bands
Figure 12.9

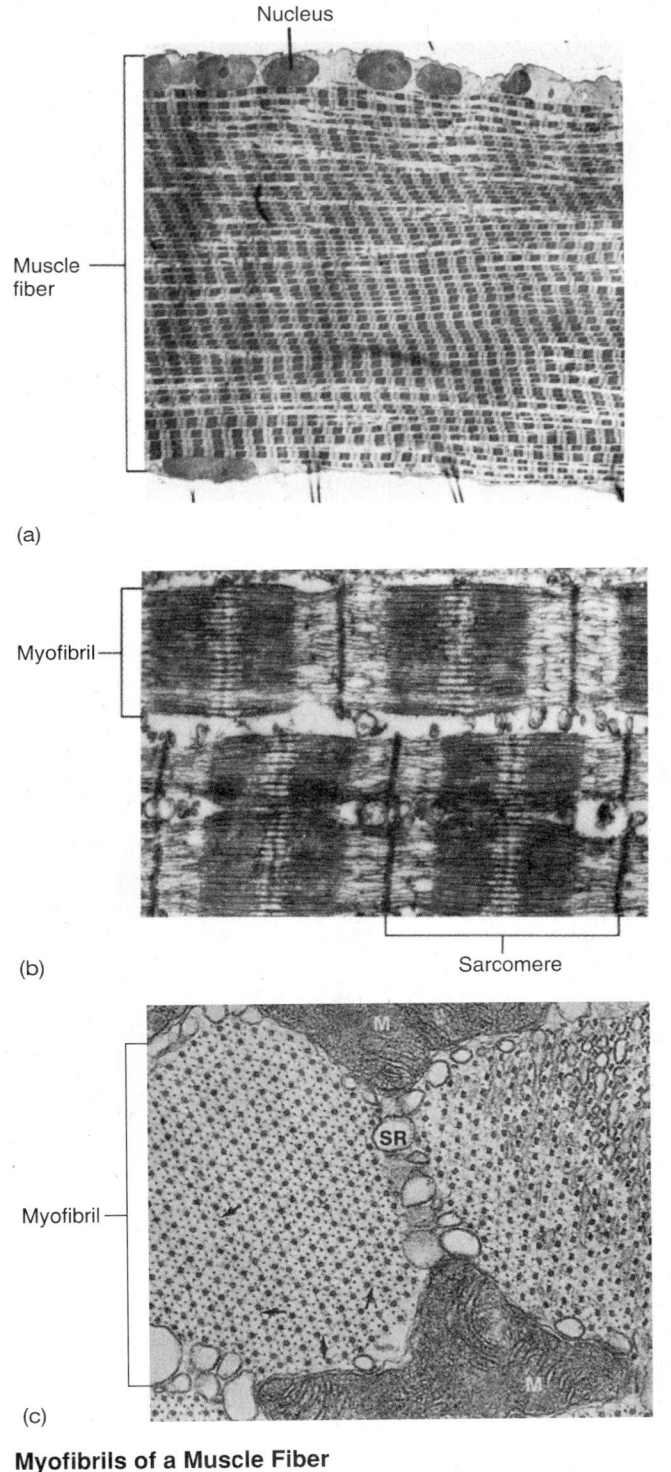

Nucleus

Muscle
fiber

(a)

Myofibril

(b)

Sarcomere

Myofibril

(c)

Myofibrils of a Muscle Fiber
Figure 12.10

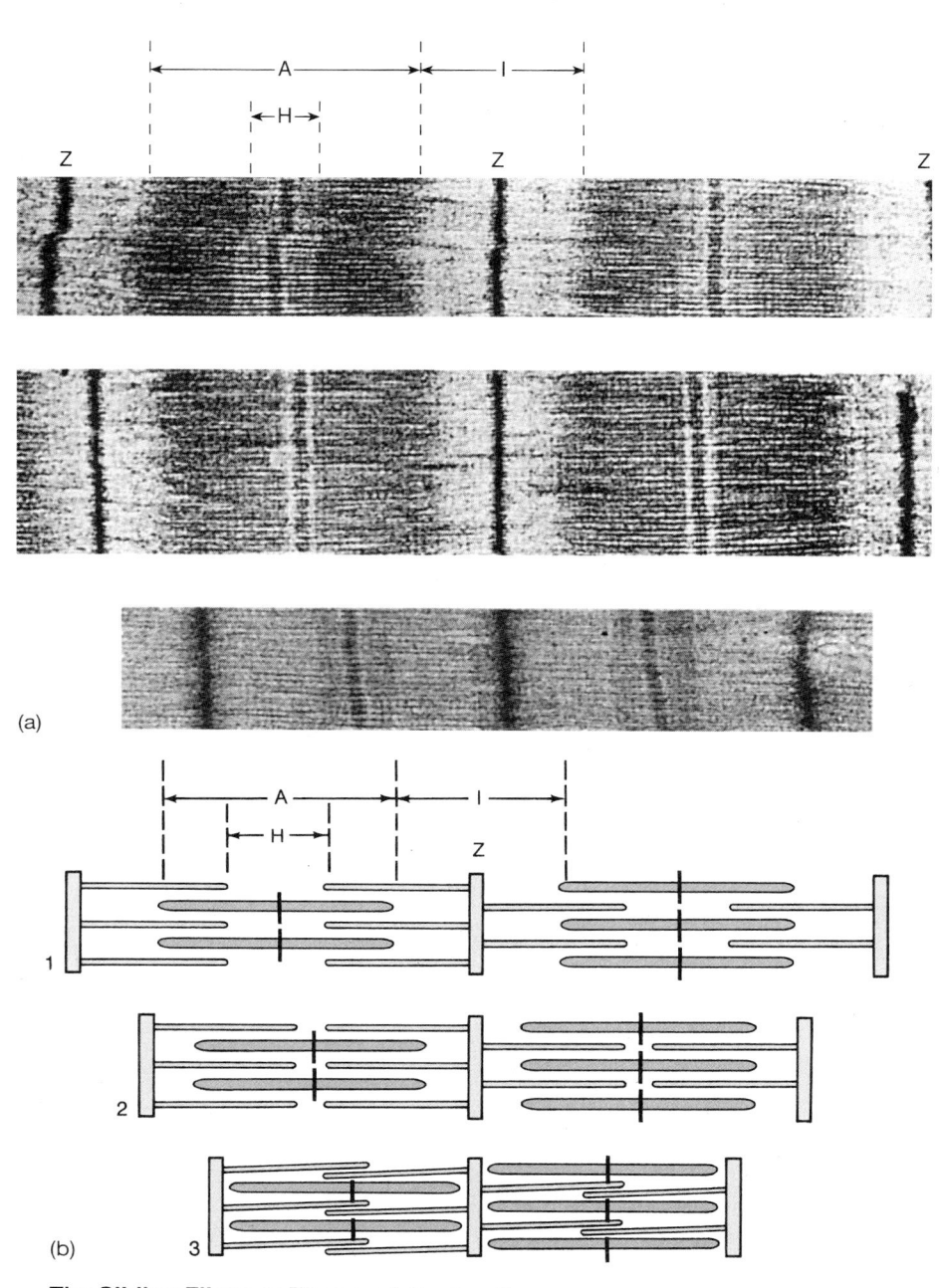

The Sliding Filament Theory of Contraction
Figure 12.11

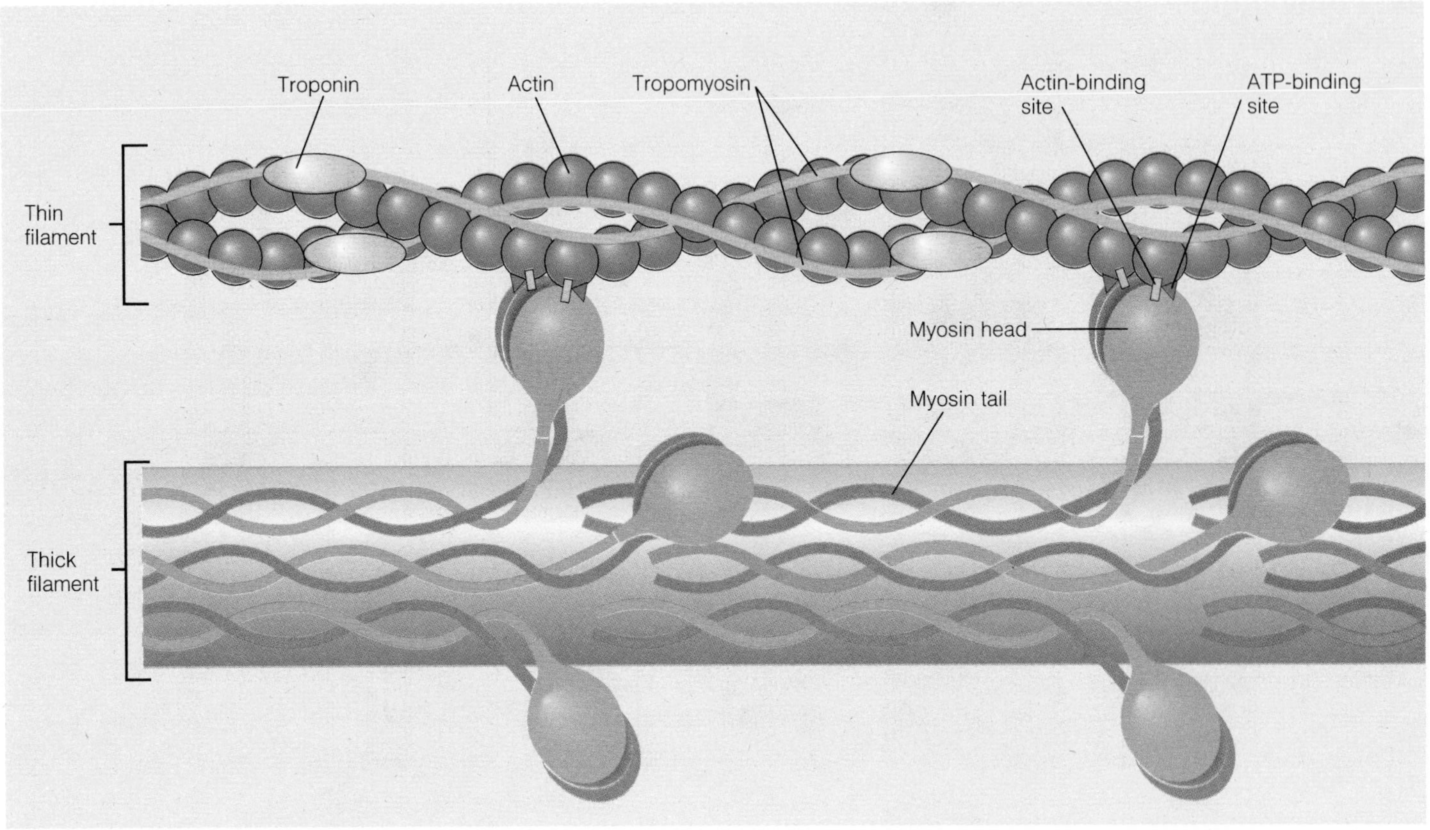
Troponin
Actin
Tropomyosin
Actin-binding site
ATP-binding site
Thin filament
Myosin head
Myosin tail
Thick filament

Bonding of Myosin to ATP and Actin
Figure 12.12

The Cross-Bridge Cycle
Figure 12.13

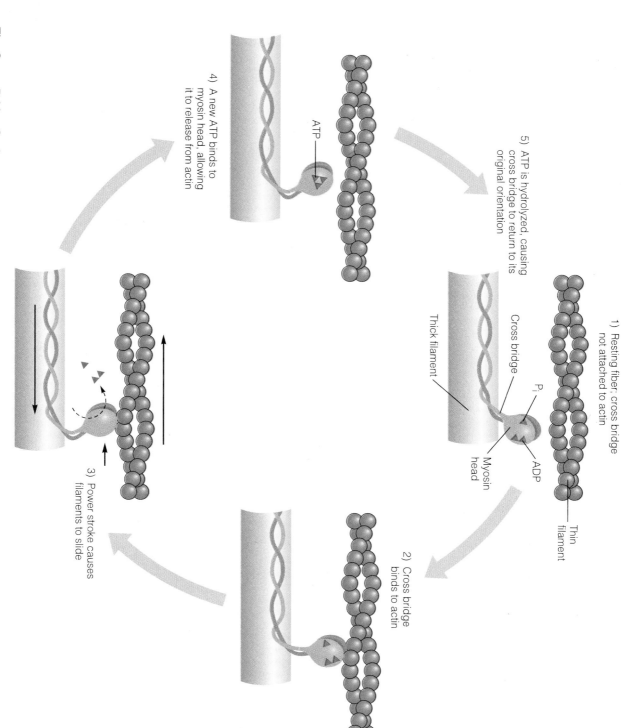

1) Resting fiber; cross bridge not attached to actin

Thin filament

2) Cross bridge binds to actin

3) Power stroke causes filaments to slide

4) A new ATP binds to myosin head, allowing it to release from actin

ATP

5) ATP is hydrolyzed, causing cross bridge to return to its original orientation

Thick filament

Cross bridge

Myosin head

P$_i$

ADP

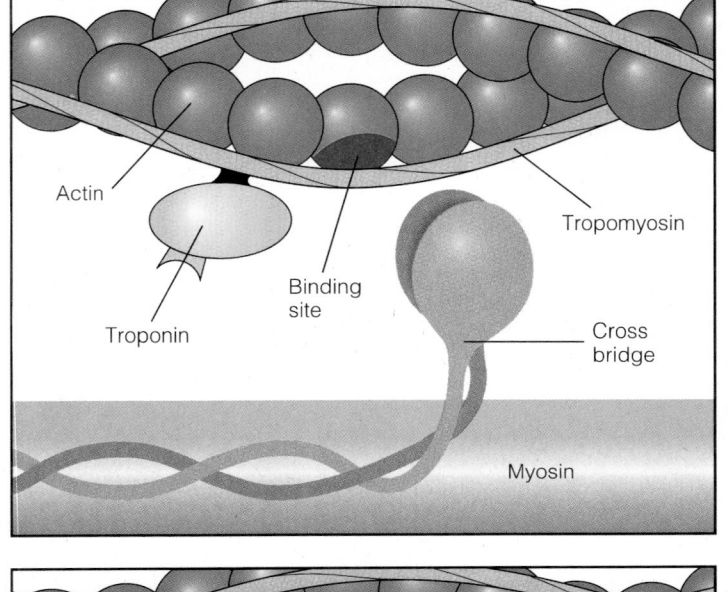

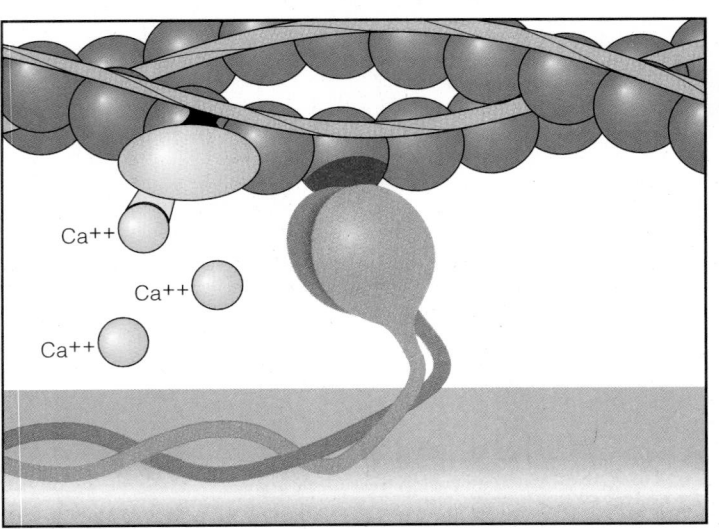

Attachment of Ca⁺⁺ to Troponin Switches on Muscle Contraction
Figure 12.15

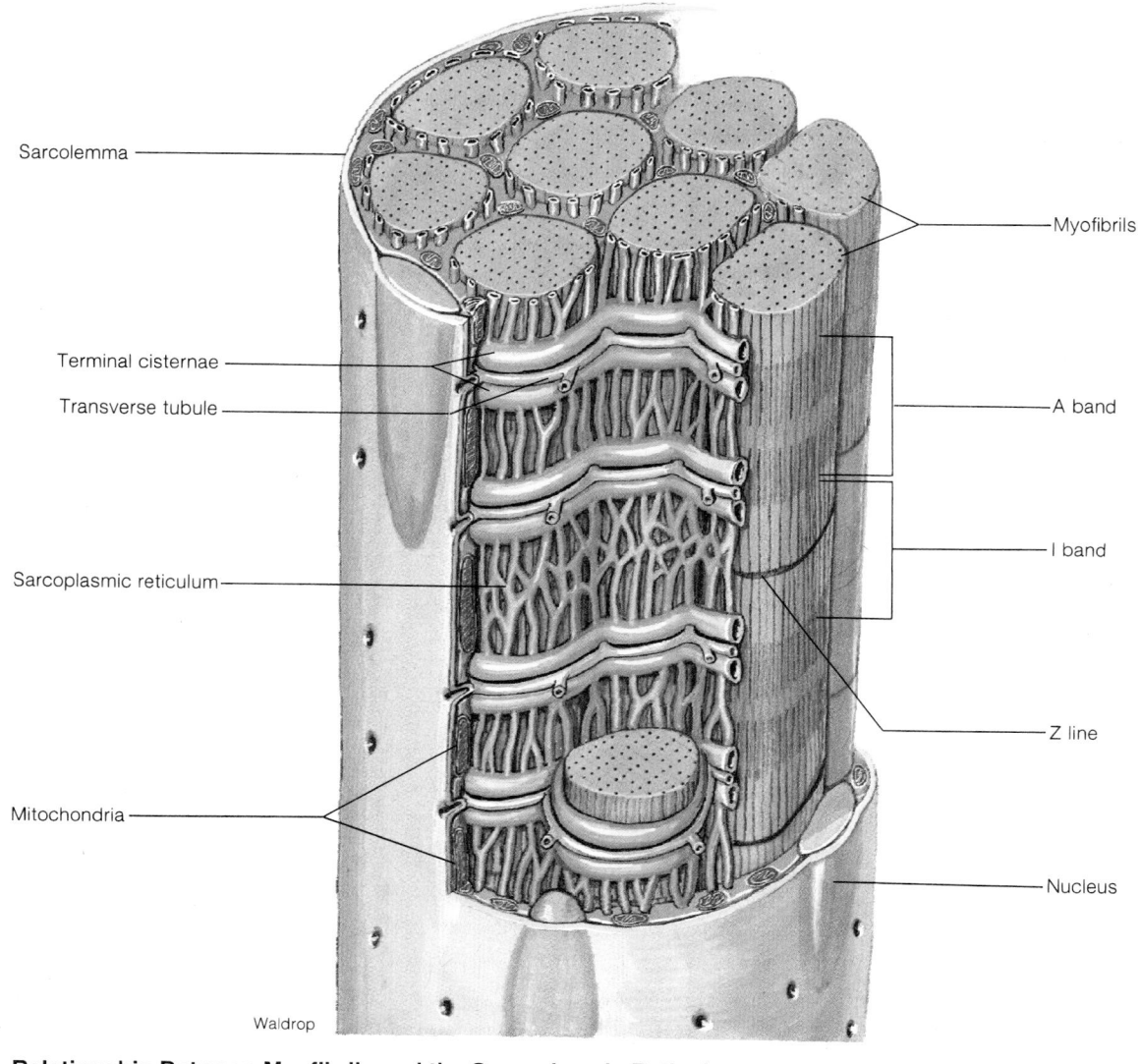

Sarcolemma

Terminal cisternae

Transverse tubule

Sarcoplasmic reticulum

Mitochondria

Myofibrils

A band

I band

Z line

Nucleus

Waldrop

Relationship Between Myofibrils and the Sarcoplasmic Reticulum
Figure 12.16

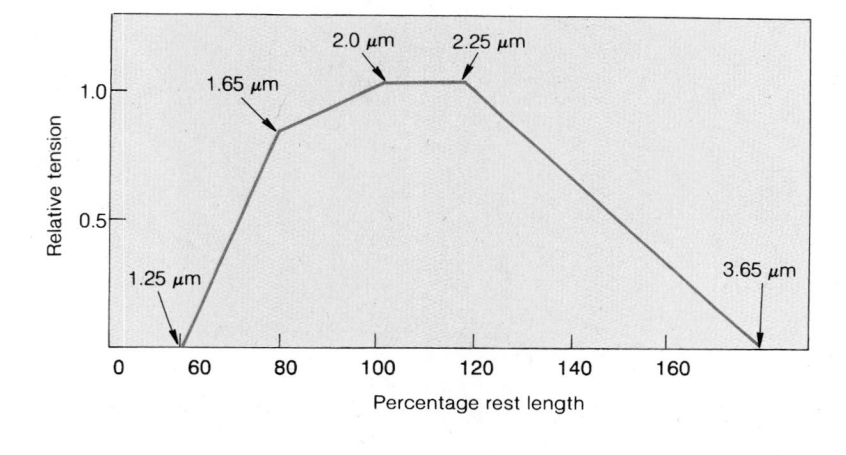

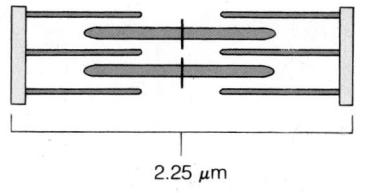

1.65 μm

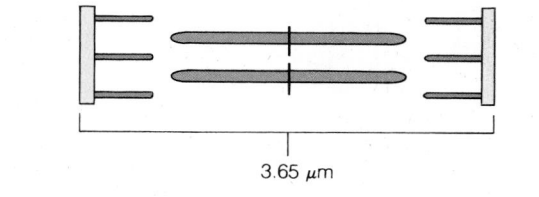

2.25 μm   3.65 μm

The Length-Tension Relationship in Skeletal Muscles
Figure 12.17

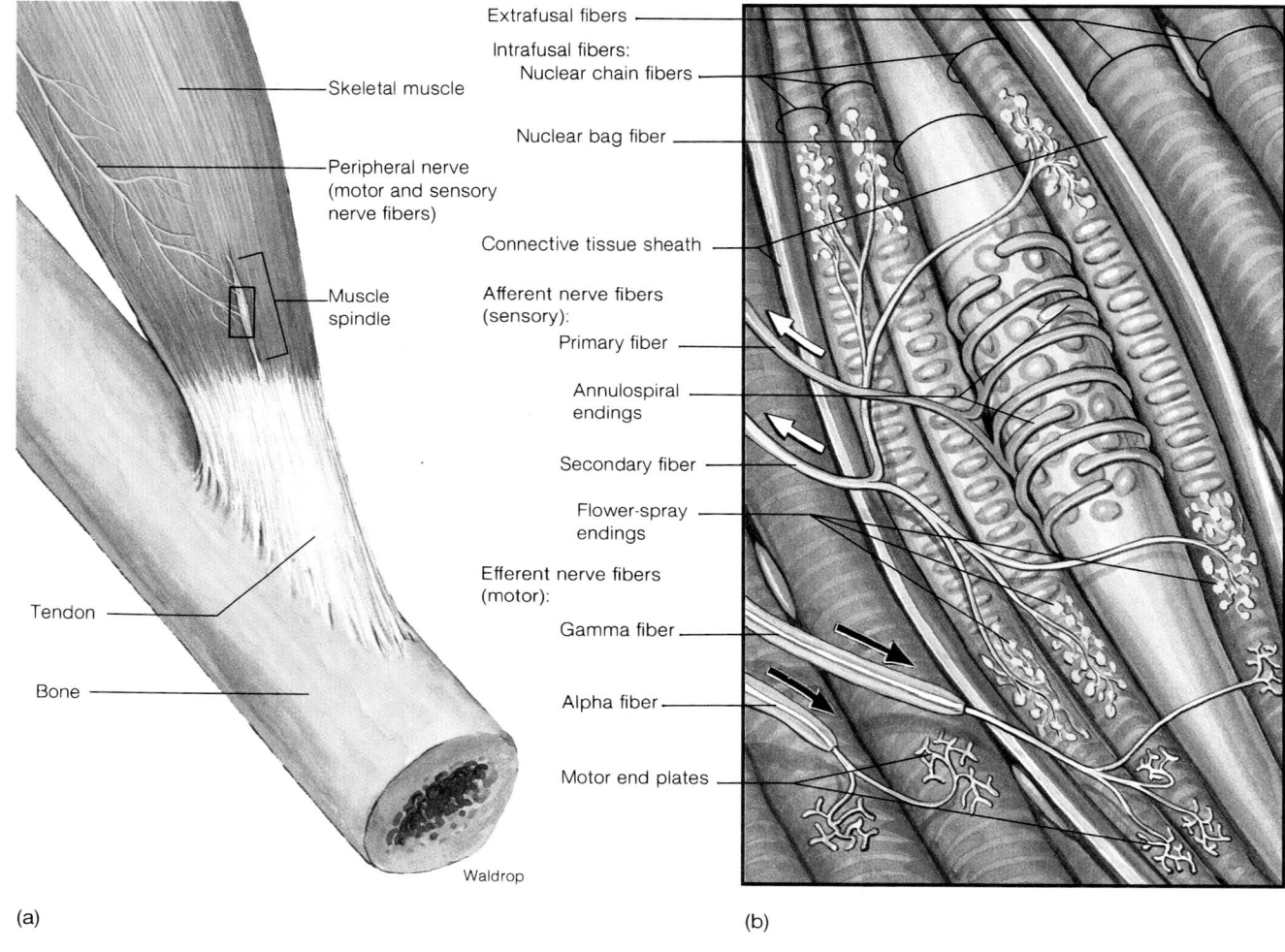

Extrafusal fibers

Intrafusal fibers:
Nuclear chain fibers

Nuclear bag fiber

Connective tissue sheath

Afferent nerve fibers
(sensory):

Primary fiber

Annulospiral
endings

Secondary fiber

Flower-spray
endings

Efferent nerve fibers
(motor):

Gamma fiber

Alpha fiber

Motor end plates

Skeletal muscle

Peripheral nerve
(motor and sensory
nerve fibers)

Muscle
spindle

Tendon

Bone

Waldrop

(a)

(b)

Structure and Innervation of Muscle Spindles
Figure 12.18

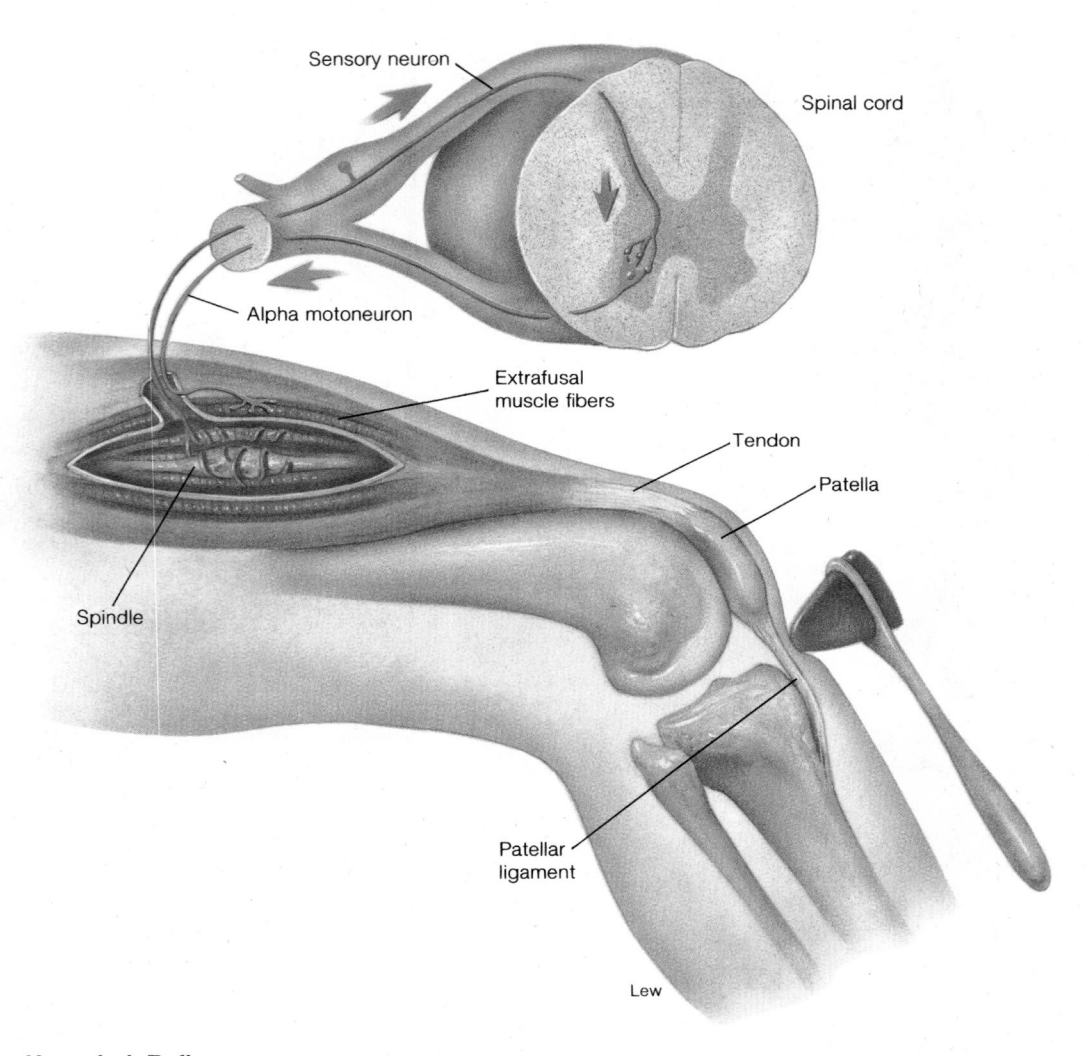

Sensory neuron

Spinal cord

Alpha motoneuron

Extrafusal
muscle fibers

Tendon

Patella

Spindle

Patellar
ligament

Lew

The Knee Jerk Reflex
Figure 12.19

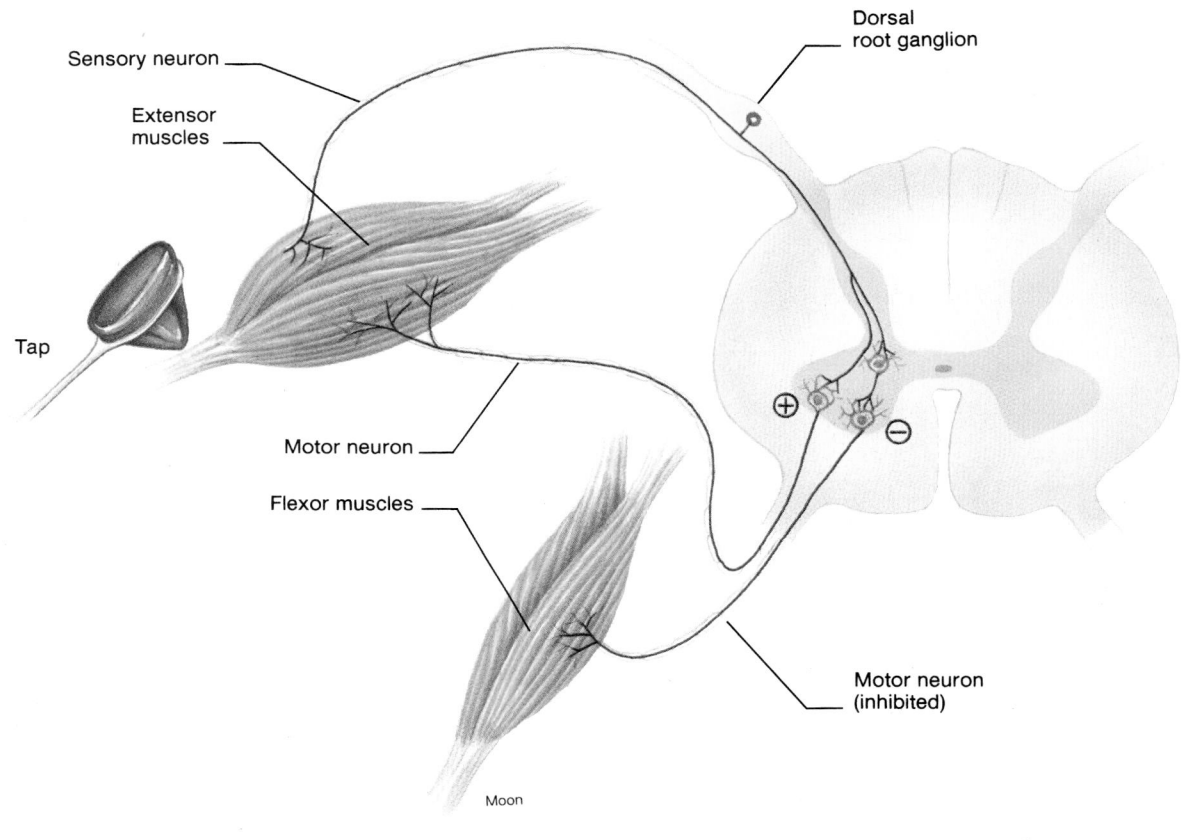

Sensory neuron

Extensor muscles

Tap

Motor neuron

Flexor muscles

Dorsal root ganglion

Motor neuron (inhibited)

Moon

A Diagram of Reciprocal Innervation
Figure 12.21

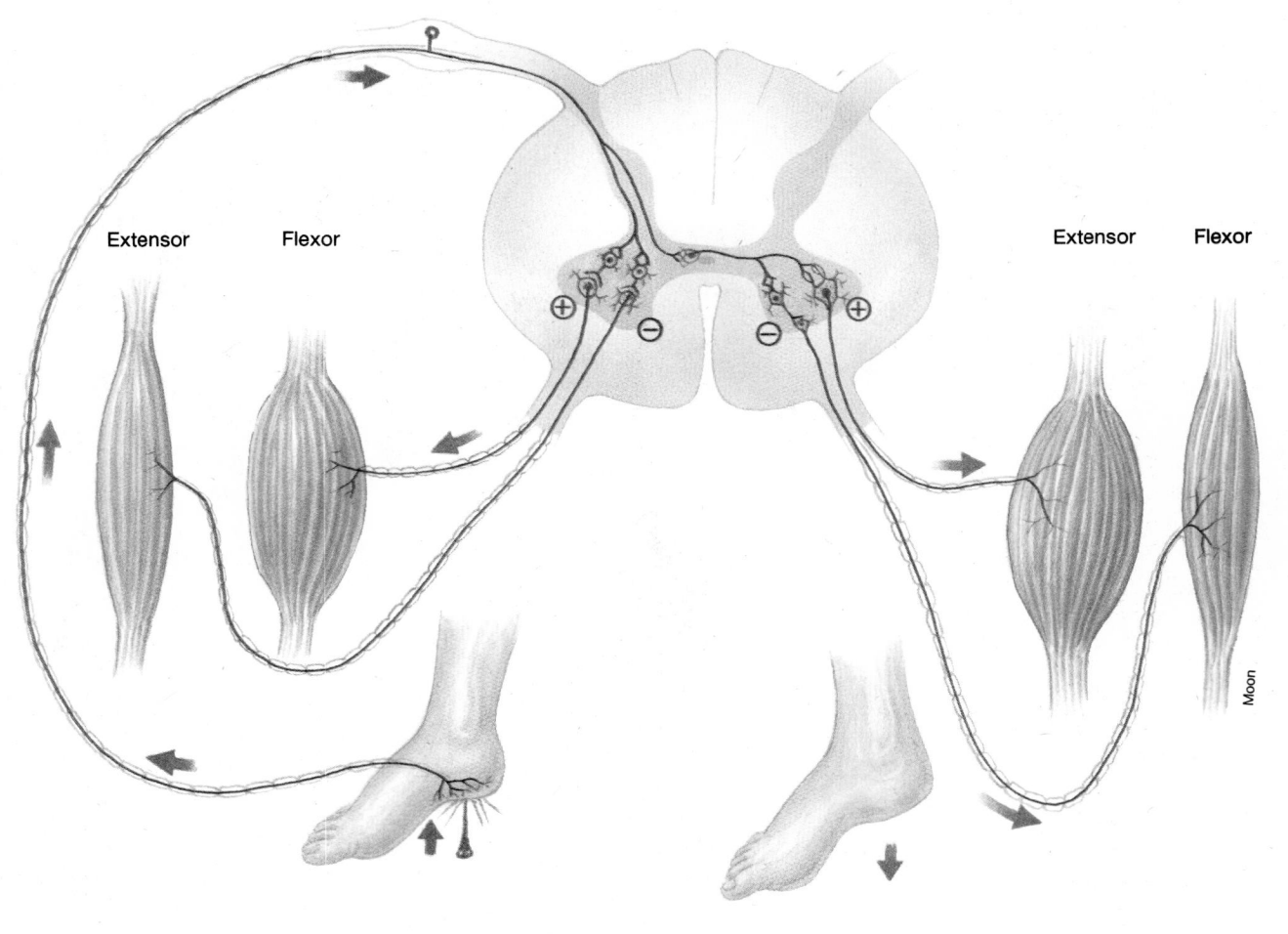

Extensor   Flexor

Extensor   Flexor

Moon

The Crossed-Extensor Reflex
Figure 12.22

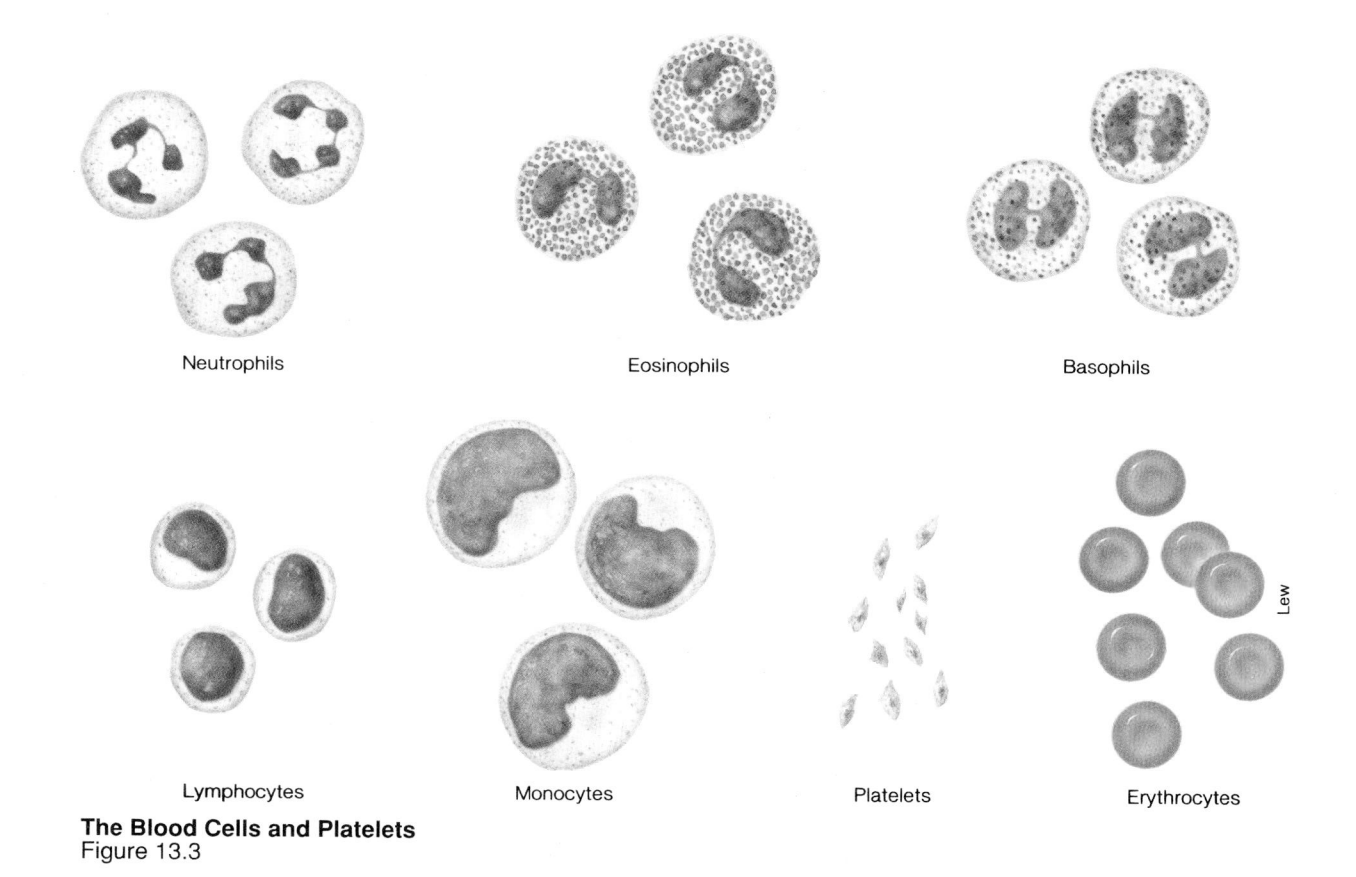

Neutrophils

Eosinophils

Basophils

Lymphocytes

Monocytes

Platelets

Erythrocytes

Lew

The Blood Cells and Platelets
Figure 13.3

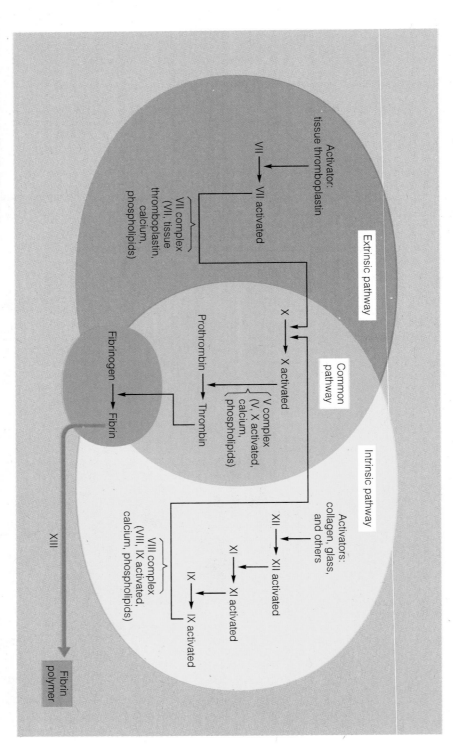

Extrinsic and Intrinsic Clotting Pathways
Figure 13.6

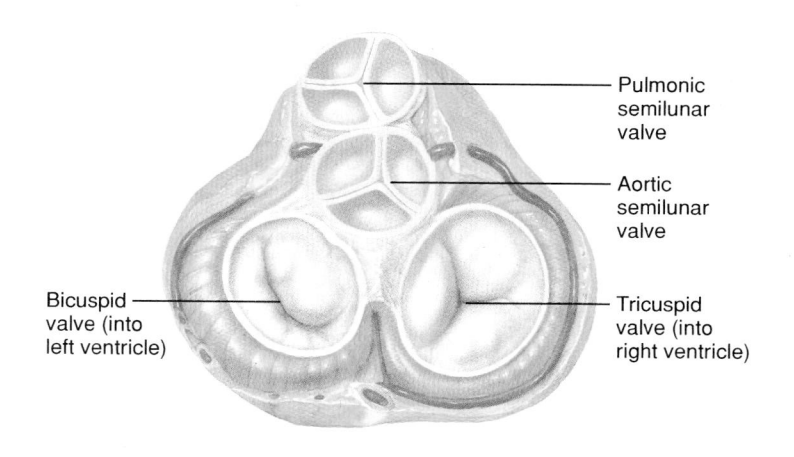

Pulmonic semilunar valve

Aortic semilunar valve

Bicuspid valve (into left ventricle)

Tricuspid valve (into right ventricle)

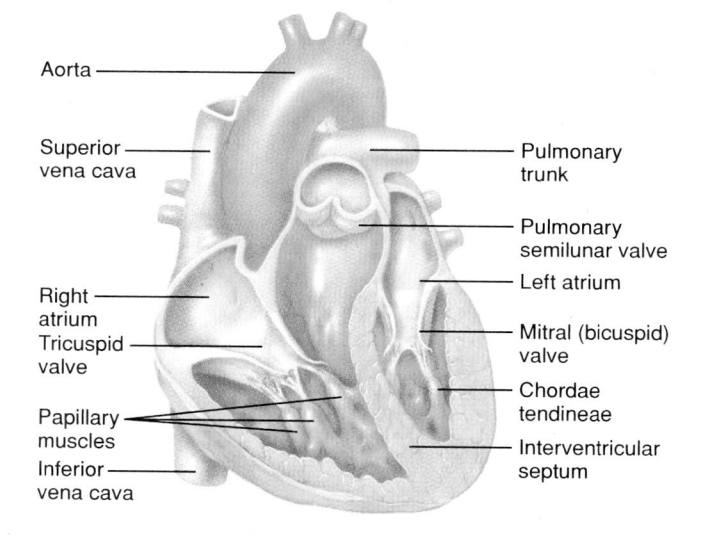

Aorta

Superior vena cava

Pulmonary trunk

Pulmonary semilunar valve

Left atrium

Right atrium

Tricuspid valve

Mitral (bicuspid) valve

Chordae tendineae

Papillary muscles

Inferior vena cava

Interventricular septum

Structure of the Heart, the AV, and Semilunar Valves
Figure 13.8

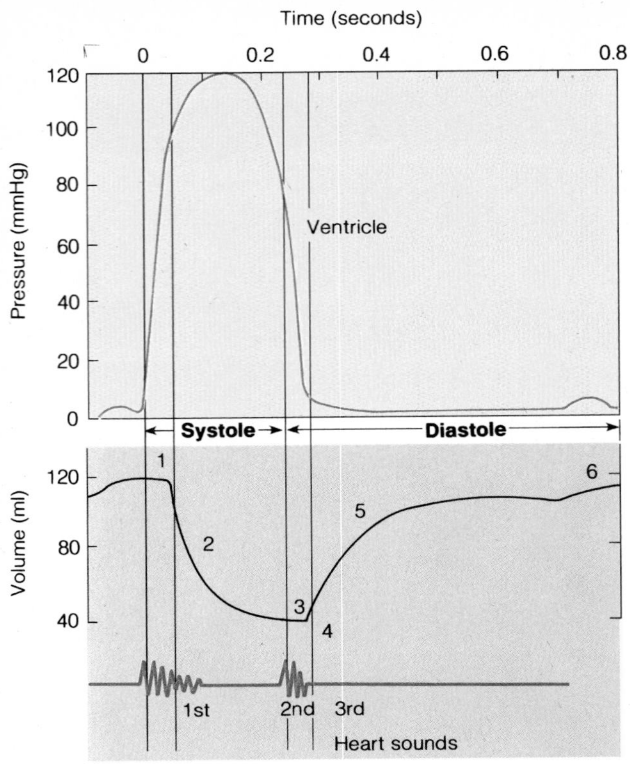

Time (seconds)

Pressure (mmHg)

Ventricle

Systole — Diastole

Volume (ml)

1
2
3
4
5
6

1st   2nd 3rd

Heart sounds

Relationship Between Intraventricular Pressure and Heart Sounds
Figure 13.11

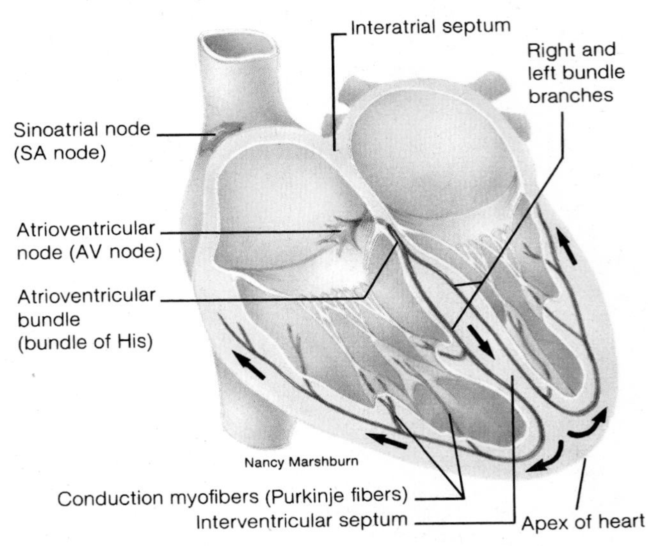

Interatrial septum

Right and left bundle branches

Sinoatrial node (SA node)

Atrioventricular node (AV node)

Atrioventricular bundle (bundle of His)

Nancy Marshburn

Conduction myofibers (Purkinje fibers)
Interventricular septum

Apex of heart

The Conduction System of the Heart
Figure 13.17

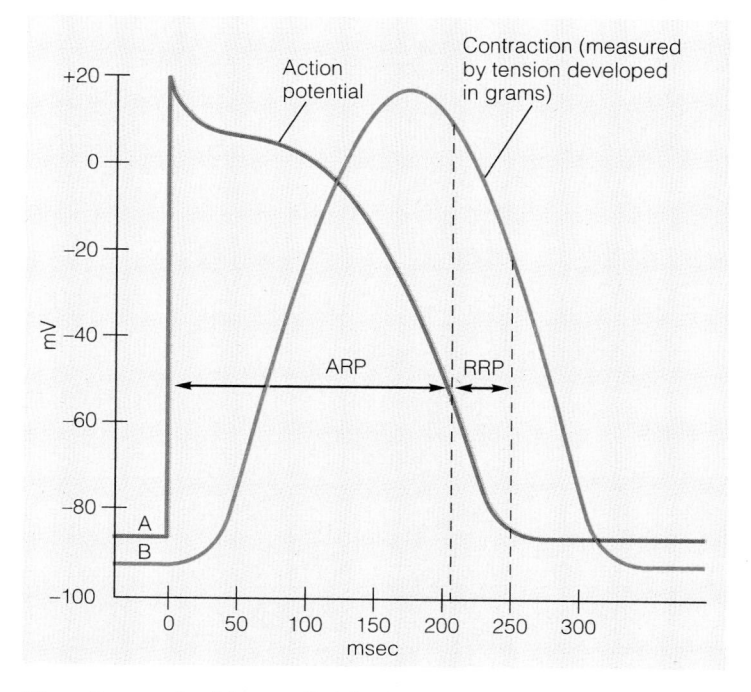

Time Course for Myocardial Action Potential and Contraction
Figure 13.18

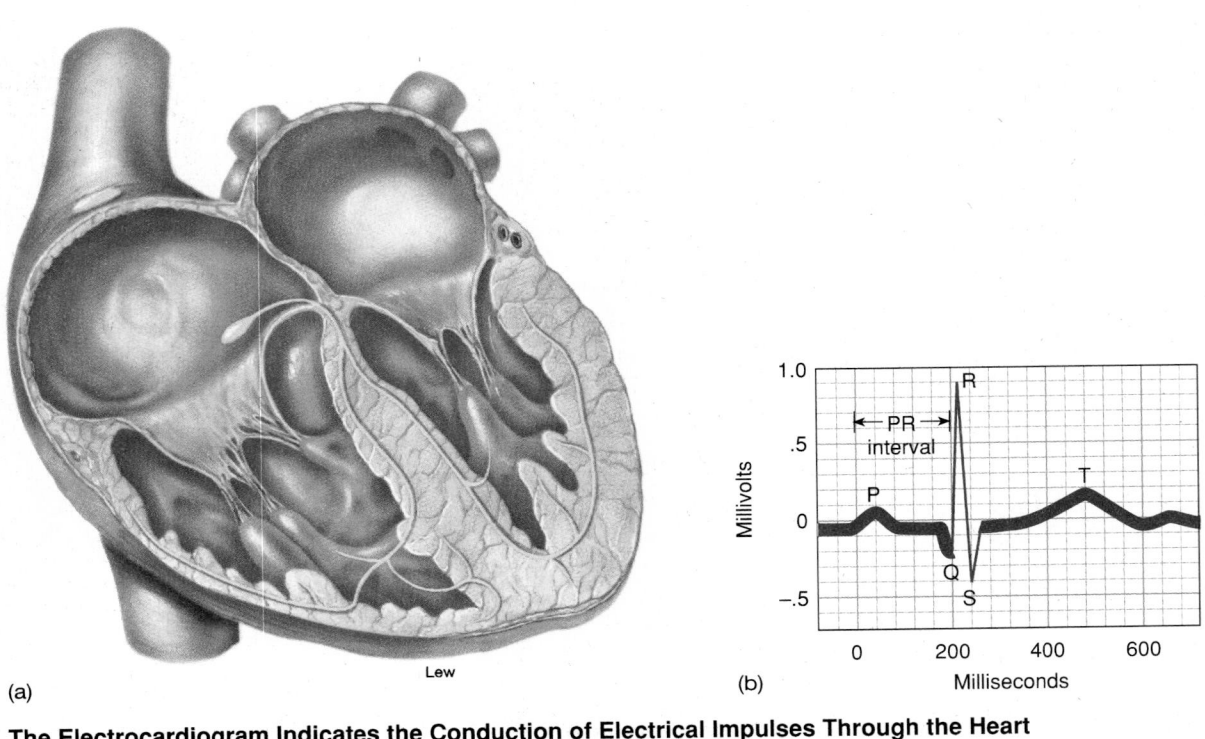

(a)

Lew

(b)

The Electrocardiogram Indicates the Conduction of Electrical Impulses Through the Heart
Figure 13.19

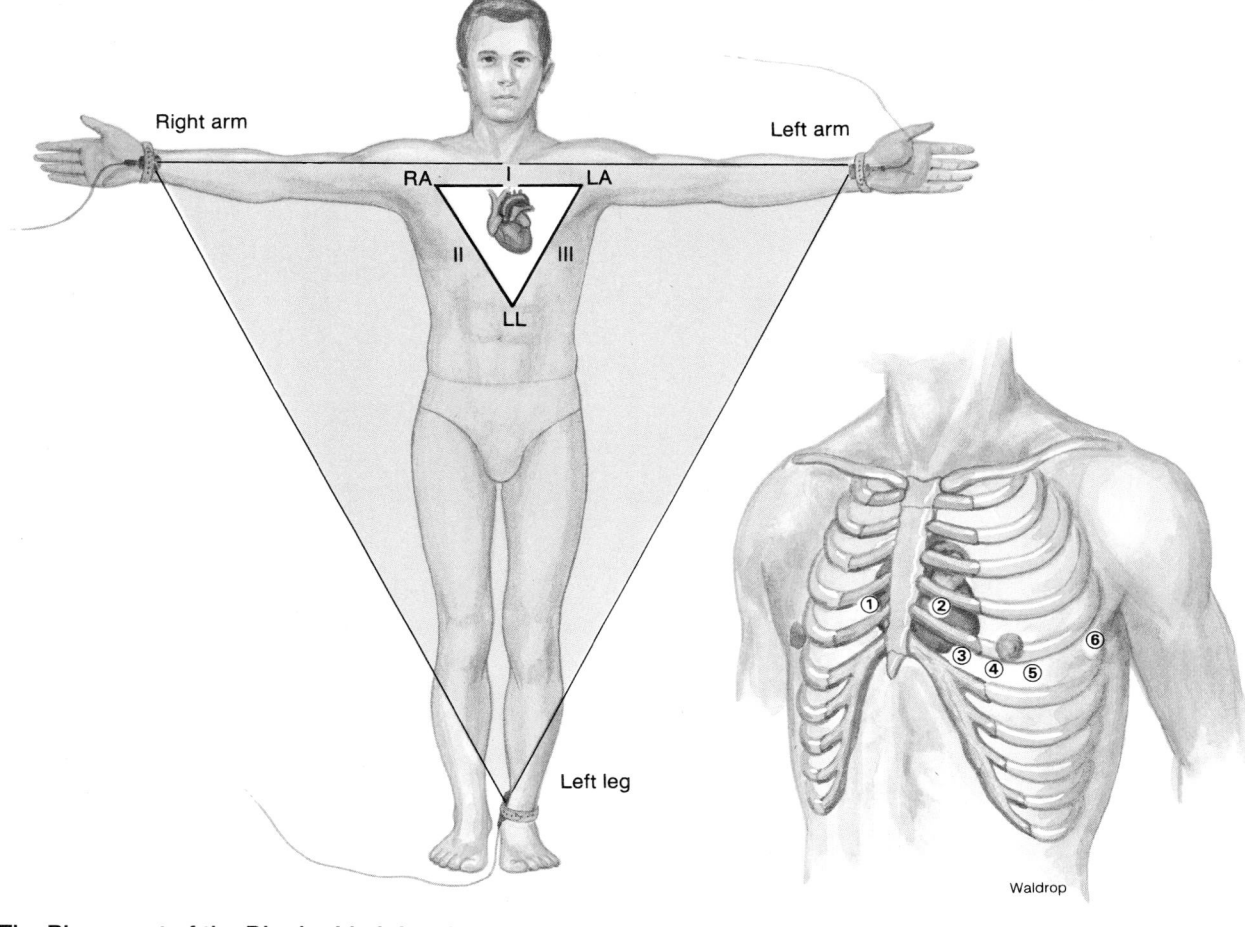

The Placement of the Bipolar Limb Leads and the Exploratory Electrode for the Unipolar Chest Leads in an Electrocardiogram
Figure 13.20

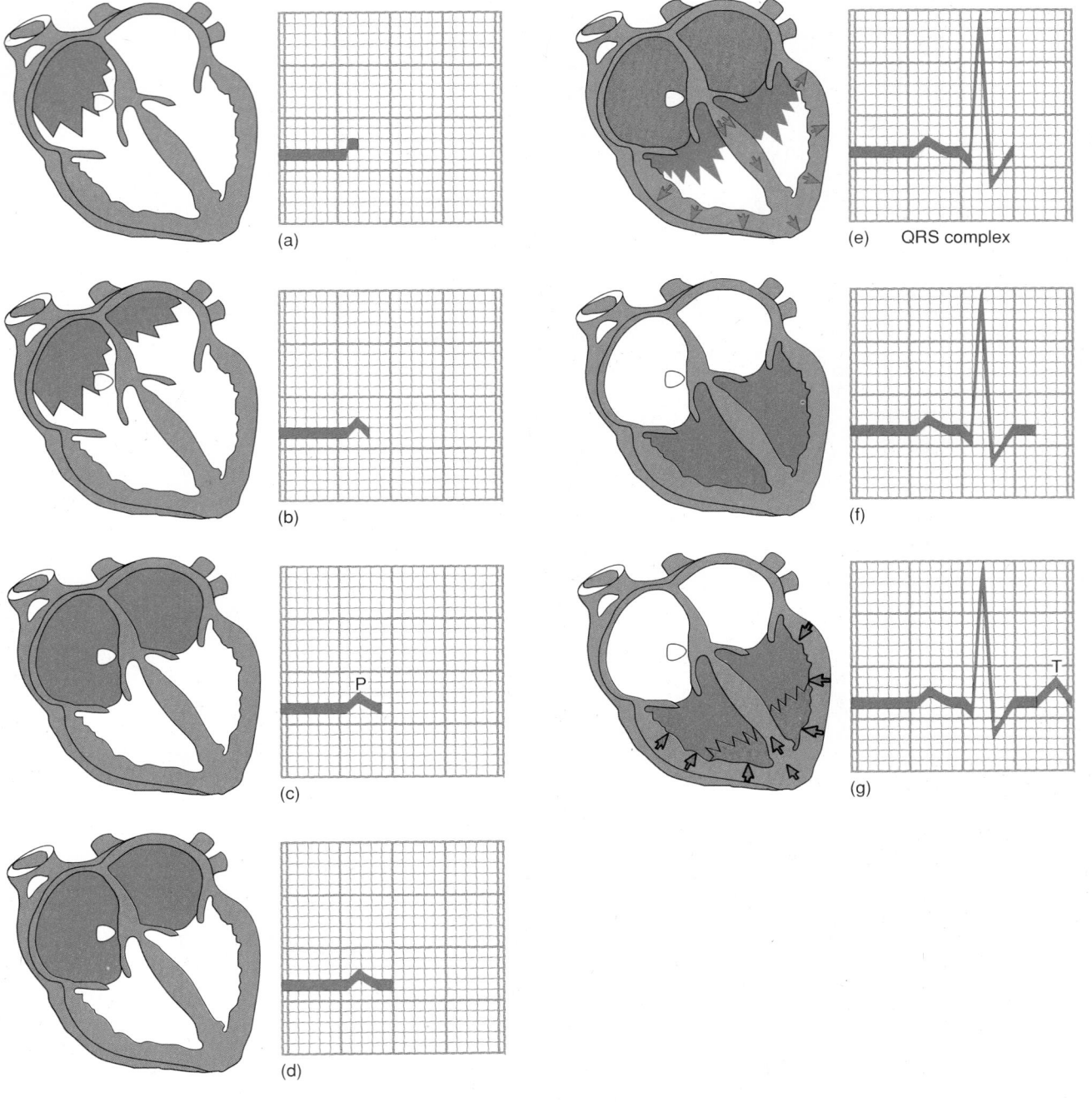

Relationship Between Impulse Conduction in the Heart and the ECG
Figure 13.21

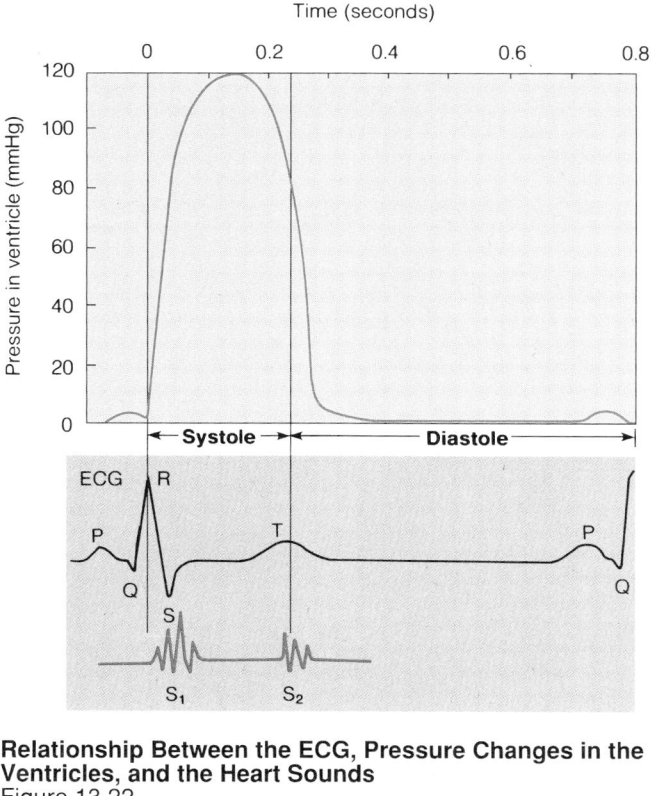

Relationship Between the ECG, Pressure Changes in the Ventricles, and the Heart Sounds
Figure 13.22

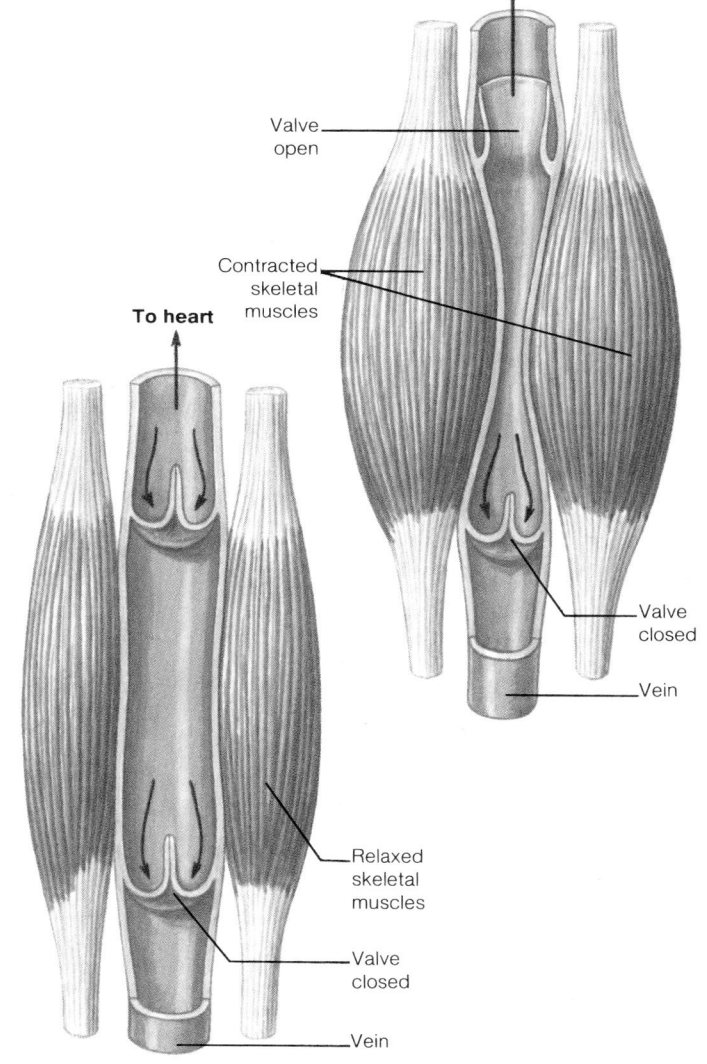

The Action of the Venous Valves
Figure 13.26

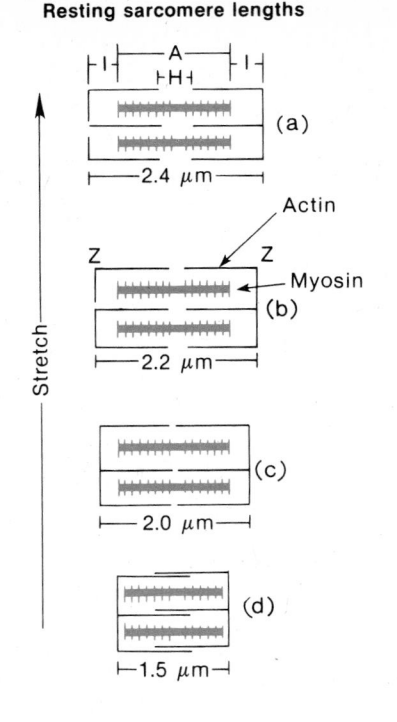

Resting sarcomere lengths

The Frank-Starling Mechanism
Figure 14.2

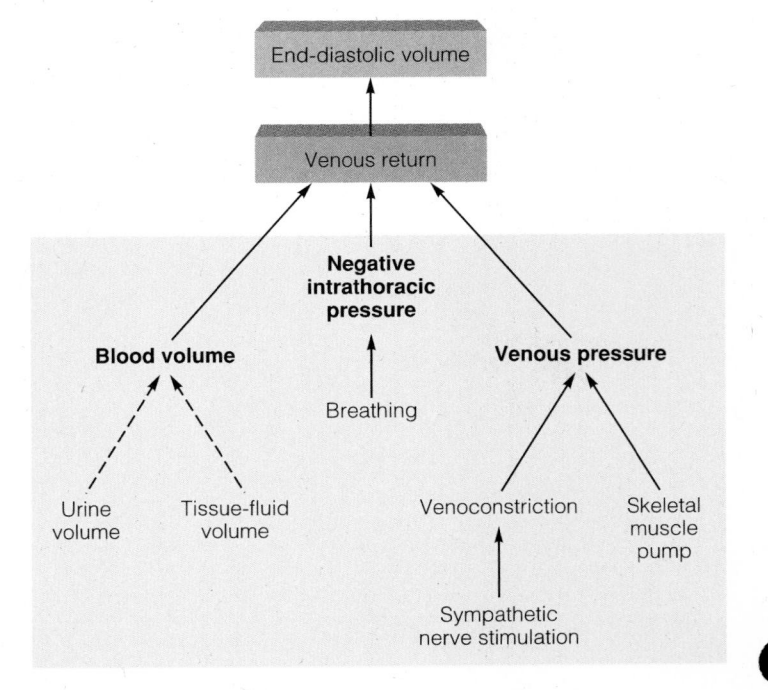

Regulation of Venous Return and End-Diastolic Volume
Figure 14.5

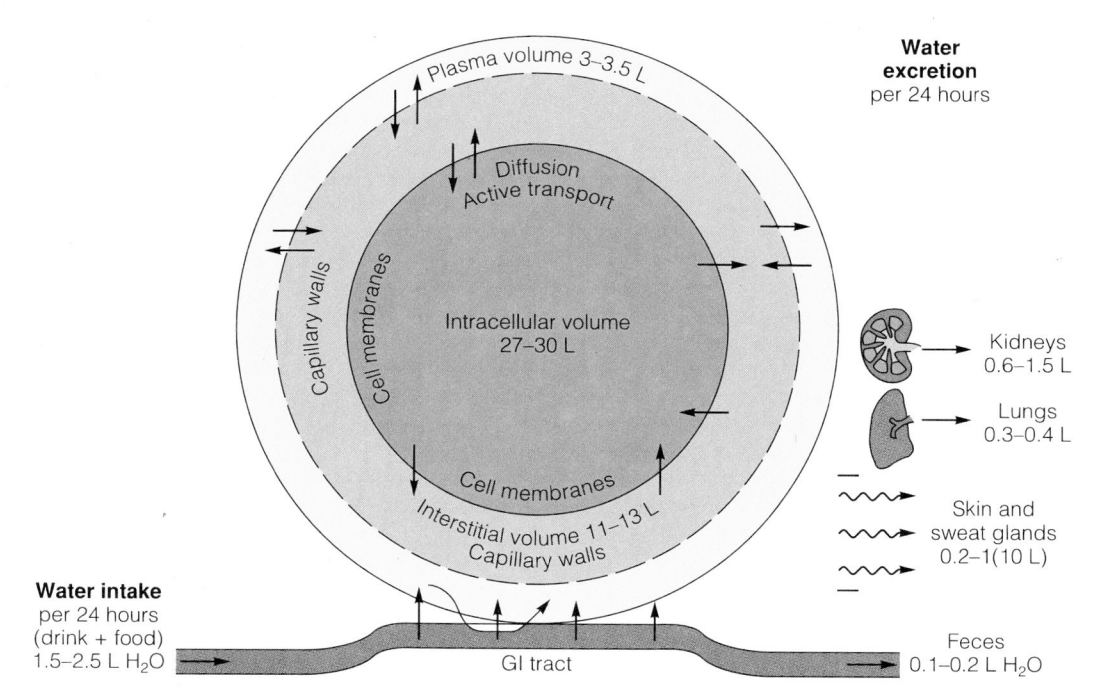

The Daily Intake and Excretion of Body Water
Figure 14.6

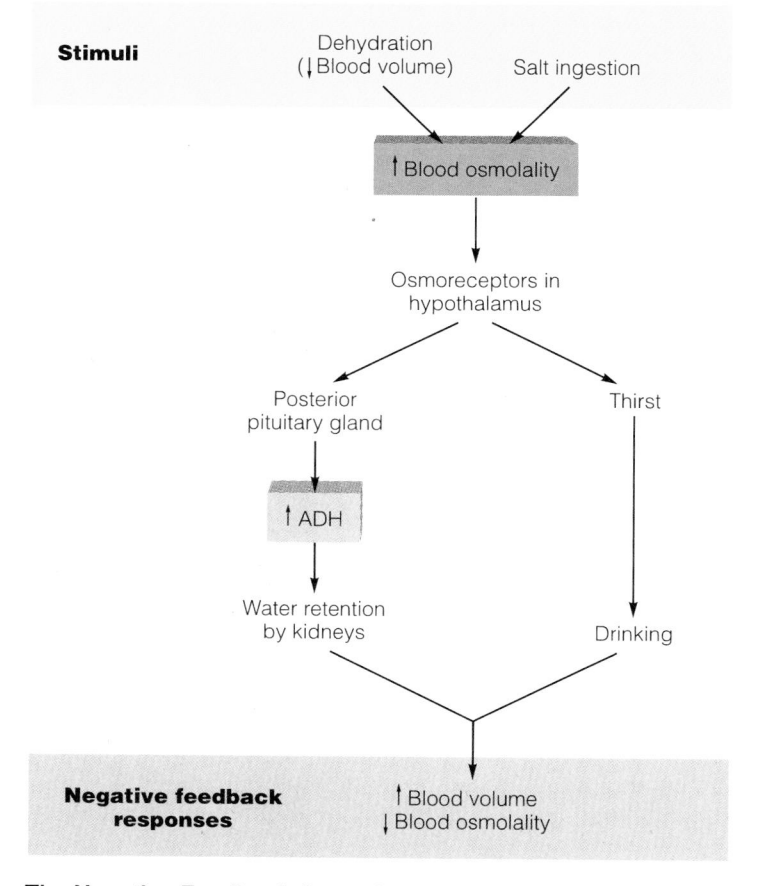

The Negative Feedback Control of Blood Volume and Blood Osmolality
Figure 14.9

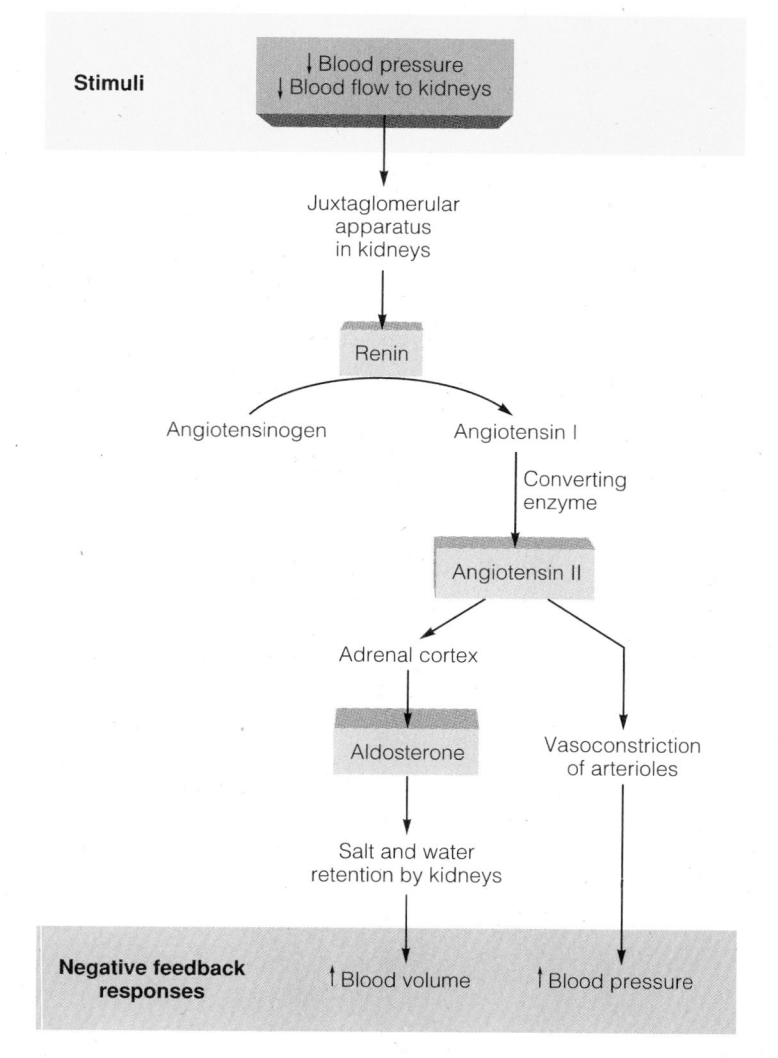

The Renin-Angiotensin-Aldosterone System
Figure 14.10

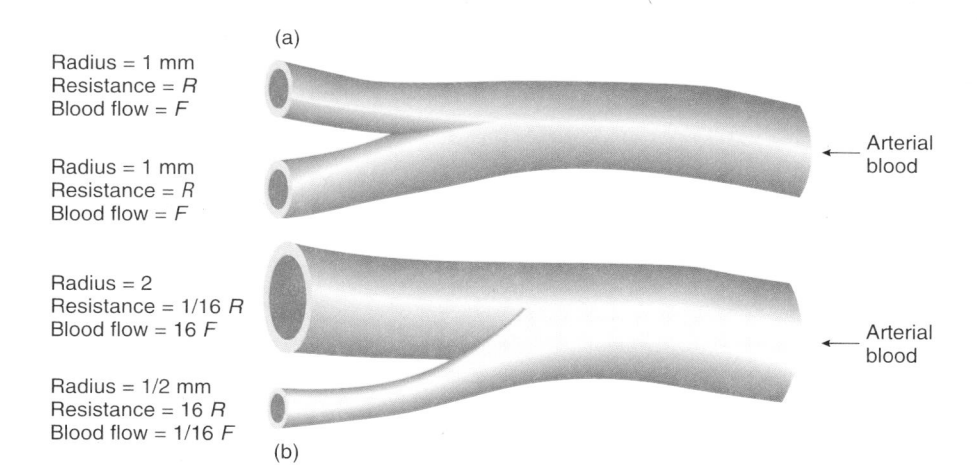

Radius = 1 mm
Resistance = R
Blood flow = F

Radius = 1 mm
Resistance = R
Blood flow = F

Radius = 2
Resistance = 1/16 R
Blood flow = 16 F

Radius = 1/2 mm
Resistance = 16 R
Blood flow = 1/16 F

(a)

(b)

Arterial blood

Arterial blood

Relationships Between Blood Flow, Vessel Radius, and Resistance
Figure 14.12

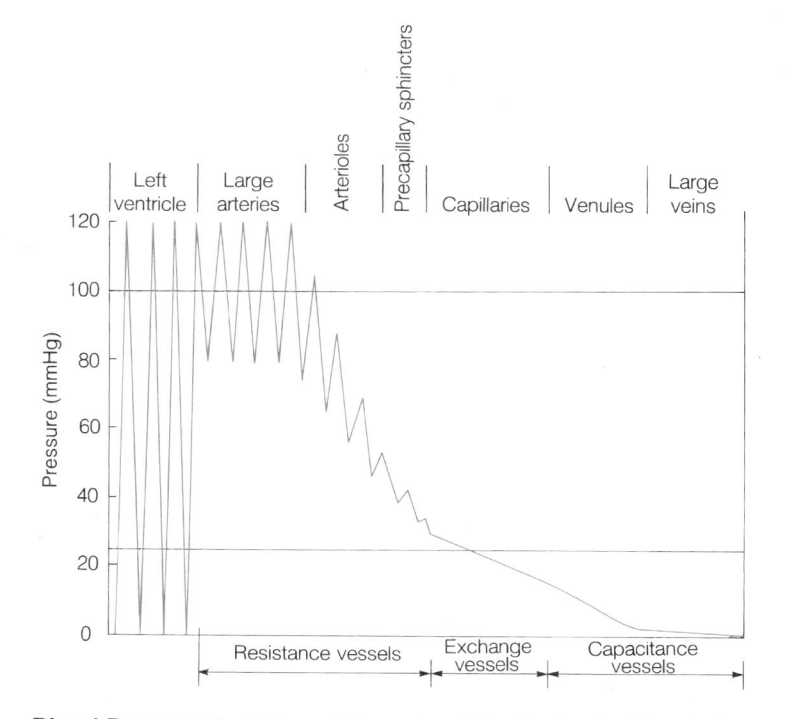

Blood Pressure in Different Vessels of the Systemic Circulation
Figure 14.13

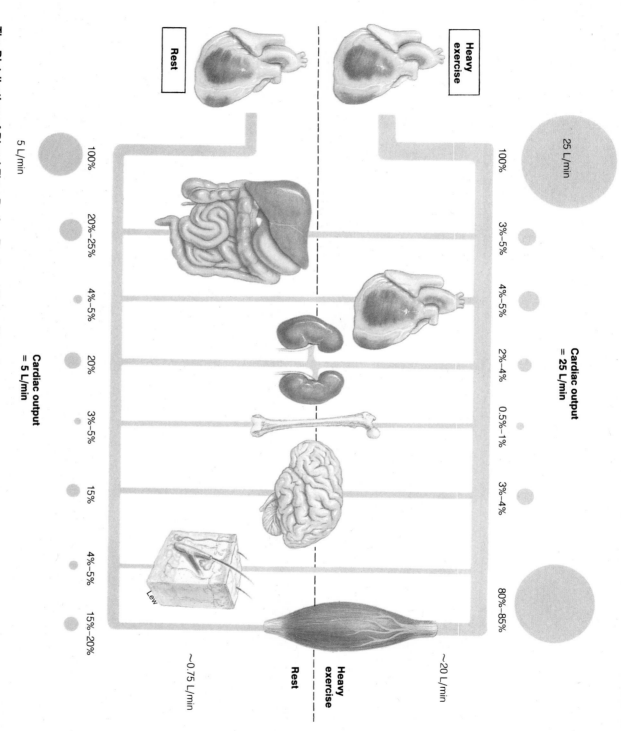

The Distribution of Blood Flow During Rest and Heavy Exercise
Figure 14.17

Heavy exercise

Rest

100%
20%–25%
4%–5%
20%
3%–5%
15%
4%–5%
15%–20%

Cardiac output = 5 L/min

5 L/min

~0.75 L/min

Rest

Heavy exercise

Cardiac output = 25 L/min

25 L/min

100%
3%–5%
4%–5%
2%–4%
0.5%–1%
3%–4%
80%–85%

~20 L/min

Lew

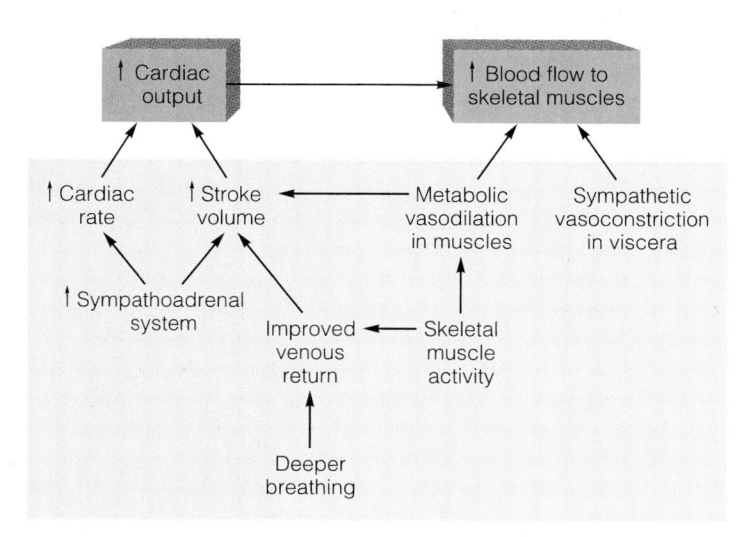

Cardiovascular Adaptations to Exercise
Figure 14.18

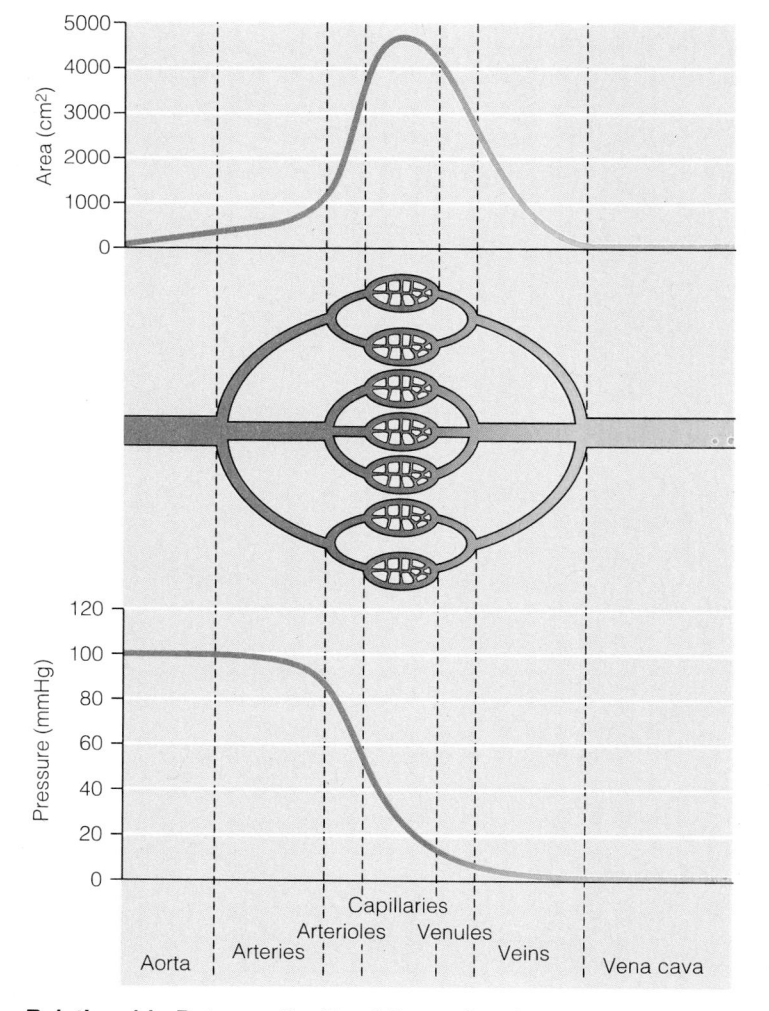

**Relationship Between the Total Cross-Sectional Area of
Blood Vessels and the Blood Pressure**
Figure 14.22

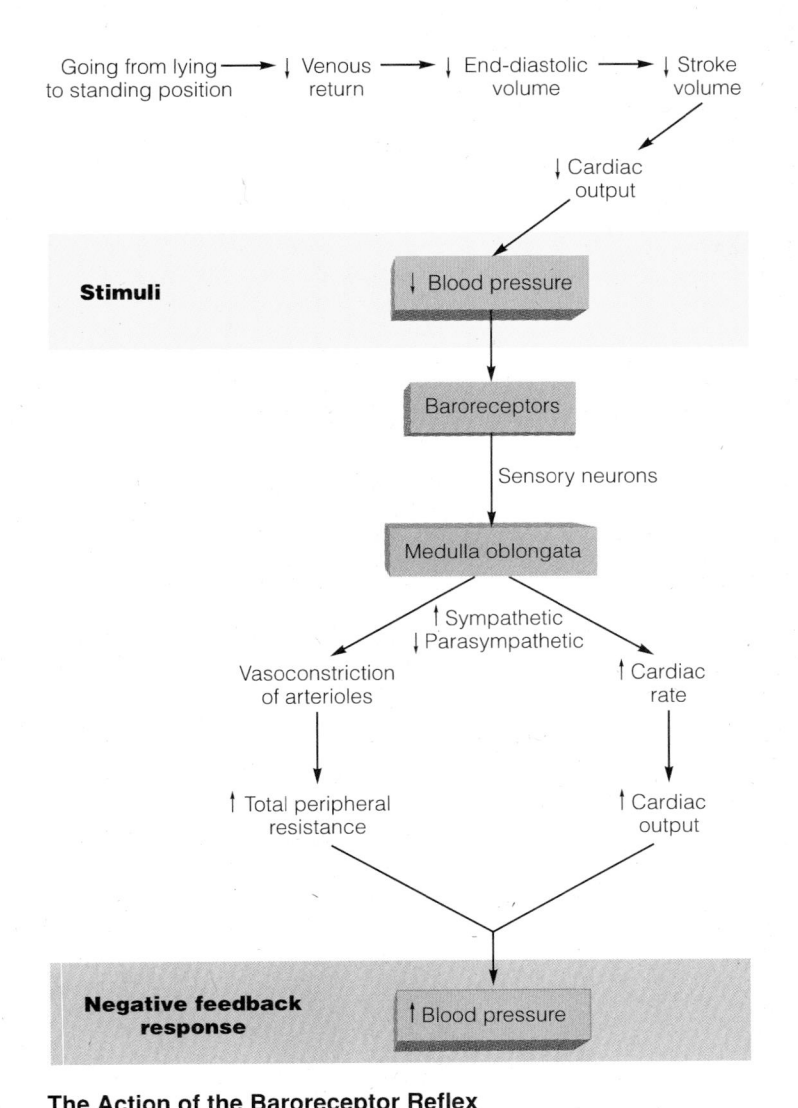

The Action of the Baroreceptor Reflex
Figure 14.25

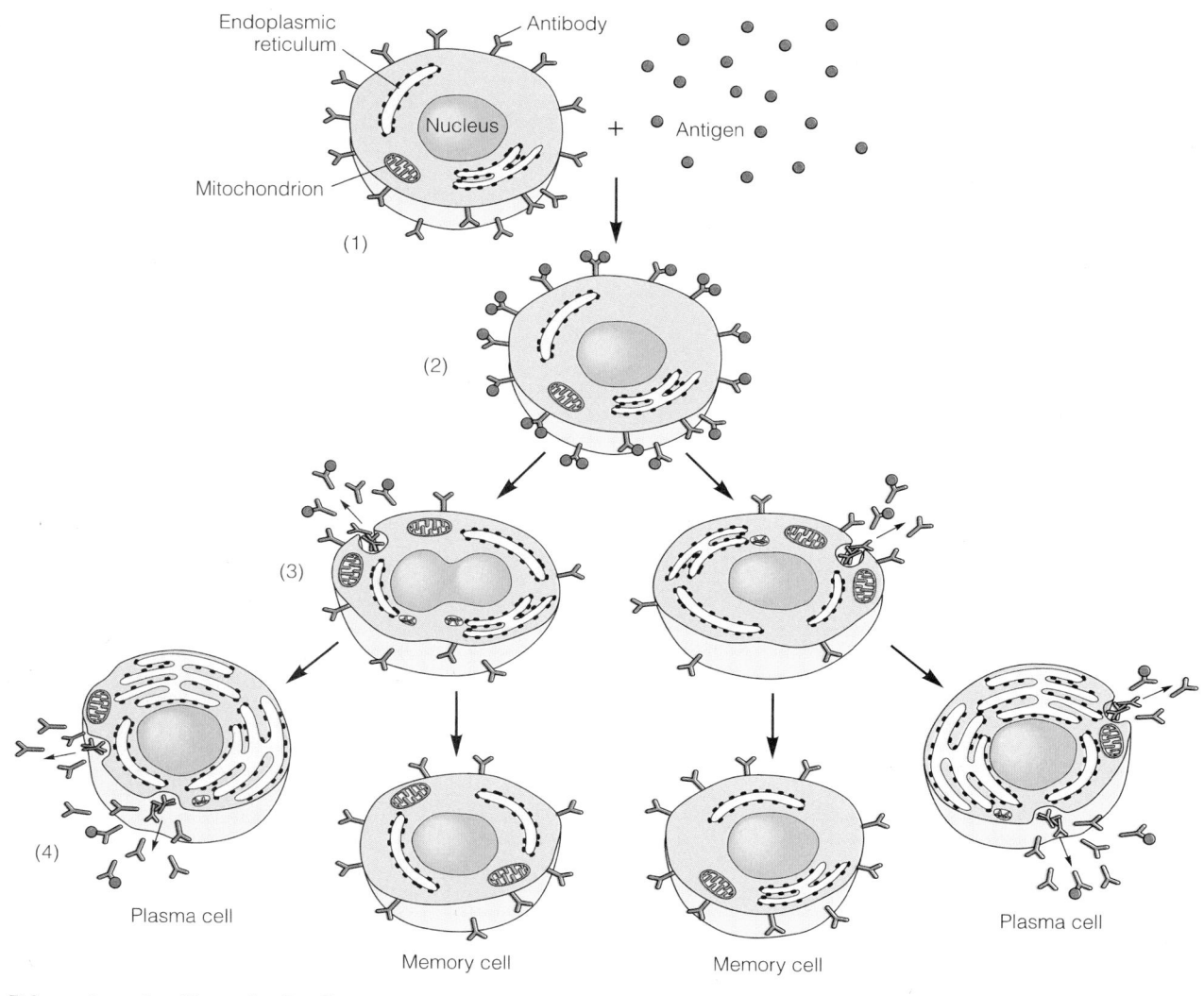

B Lymphocytes Have Antibodies on Their Surface That Function as Receptors for Specific Antigens
Figure 15.5

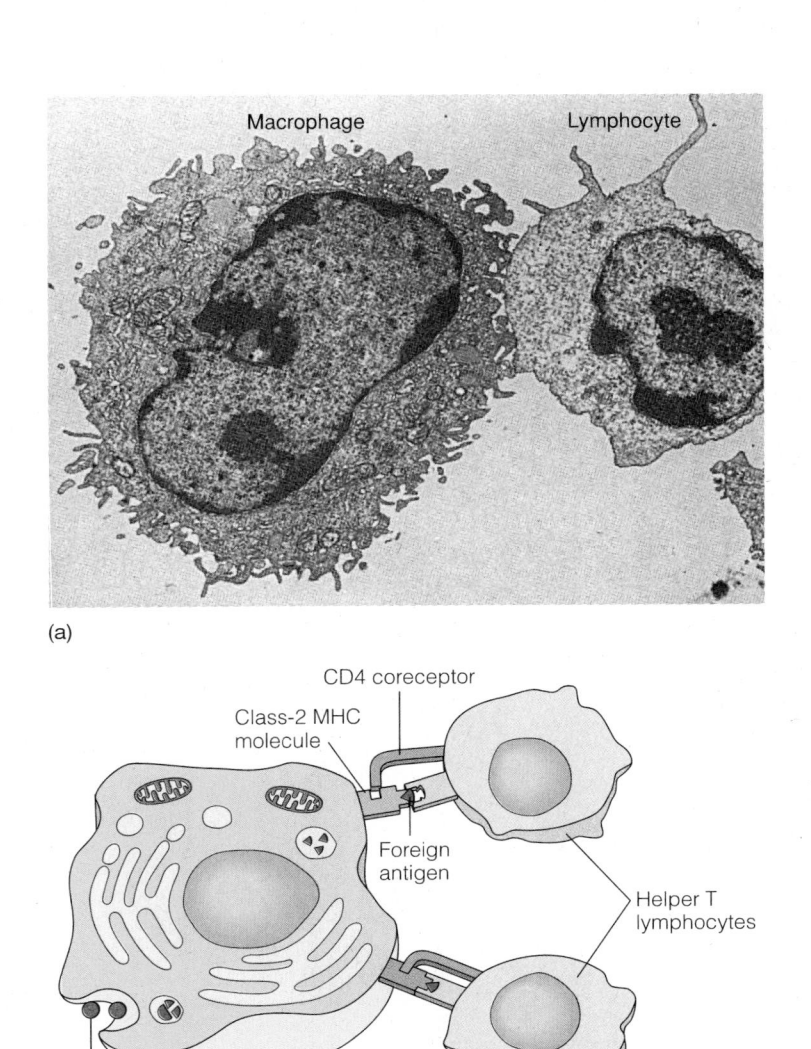

Macrophage Lymphocyte

(a)

CD4 coreceptor

Class-2 MHC
molecule

Foreign
antigen

Helper T
lymphocytes

Foreign
particle

Macrophage

Activated
helper T cell

B lymphocyte

(b)

A Macrophage Presents Antigens to Helper T Lymphocytes
Figure 15.18

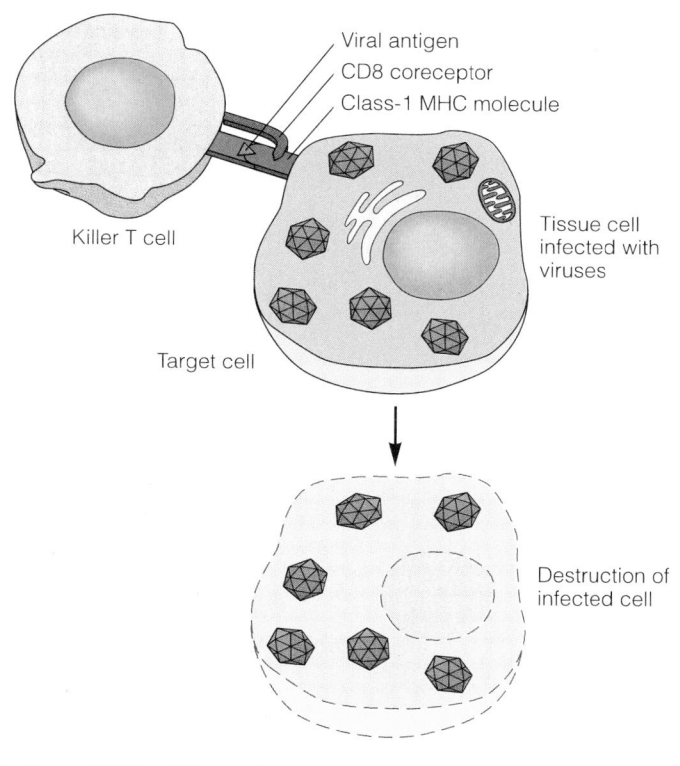

Viral antigen
CD8 coreceptor
Class-1 MHC molecule

Killer T cell

Tissue cell
infected with
viruses

Target cell

Destruction of
infected cell

Action of Killer T Lymphocytes
Figure 15.19

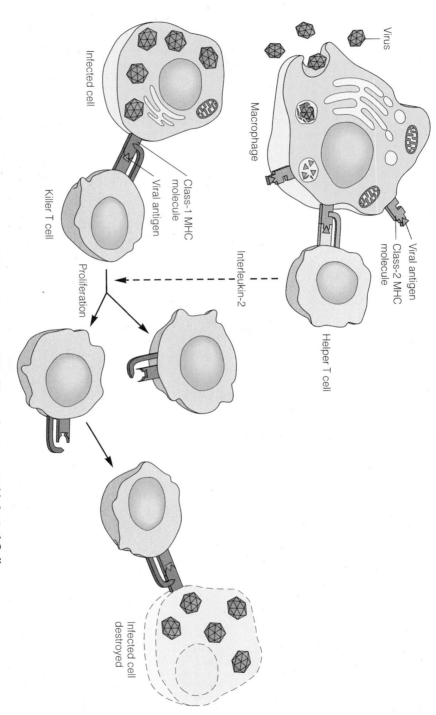

Interaction Between Macrophages, Helper T Lymphocytes, Killer T Lymphoctyes, and Infected Cells
Figure 15.20

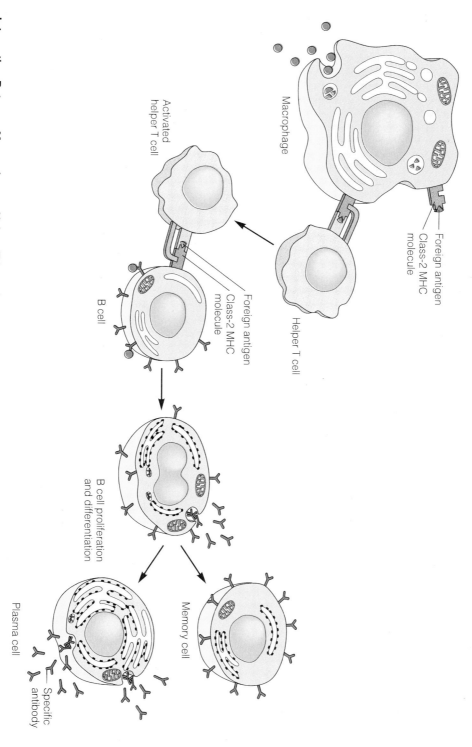

Macrophage

Foreign antigen
Class-2 MHC
molecule

Helper T cell

Activated
helper T cell

B cell

Foreign antigen
Class-2 MHC
molecule

B cell proliferation
and differentiation

Memory cell

Plasma cell

Specific
antibody

Interaction Between Macrophages, Helper T Lymphocytes, and B Lymphocytes
Figure 15.21

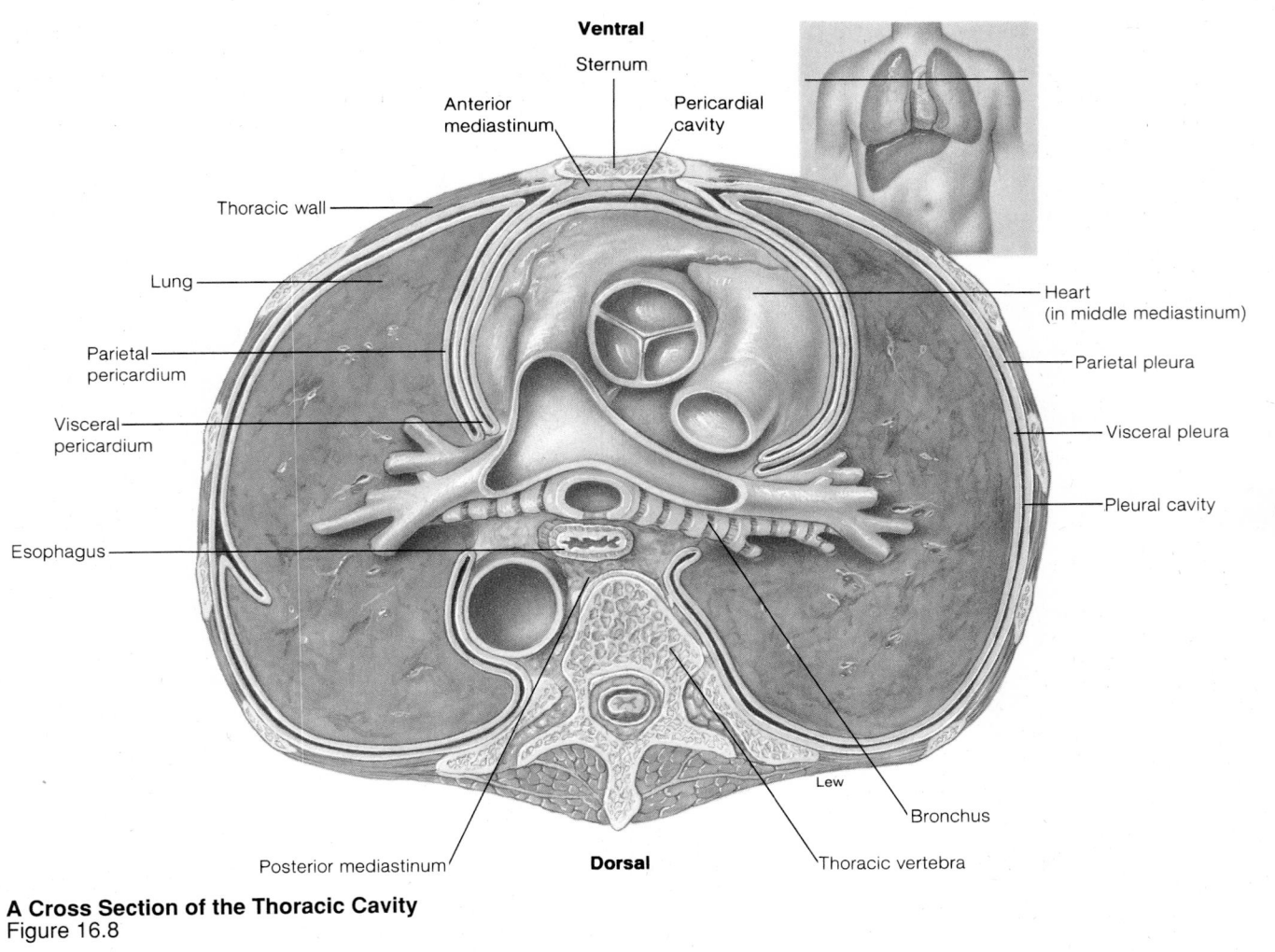

Ventral

Sternum

Anterior mediastinum

Pericardial cavity

Thoracic wall

Lung

Parietal pericardium

Visceral pericardium

Esophagus

Heart (in middle mediastinum)

Parietal pleura

Visceral pleura

Pleural cavity

Lew

Bronchus

Posterior mediastinum

Dorsal

Thoracic vertebra

A Cross Section of the Thoracic Cavity
Figure 16.8

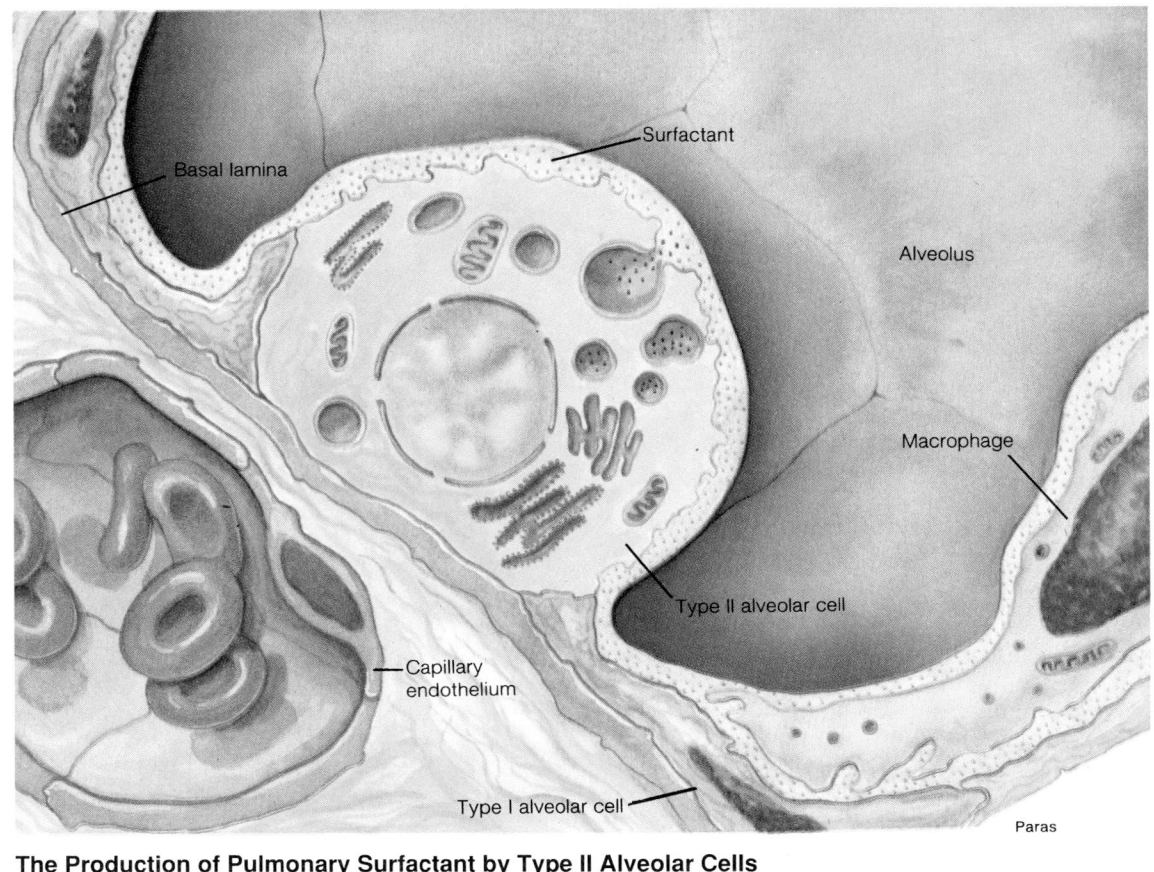

The Production of Pulmonary Surfactant by Type II Alveolar Cells
Figure 16.12

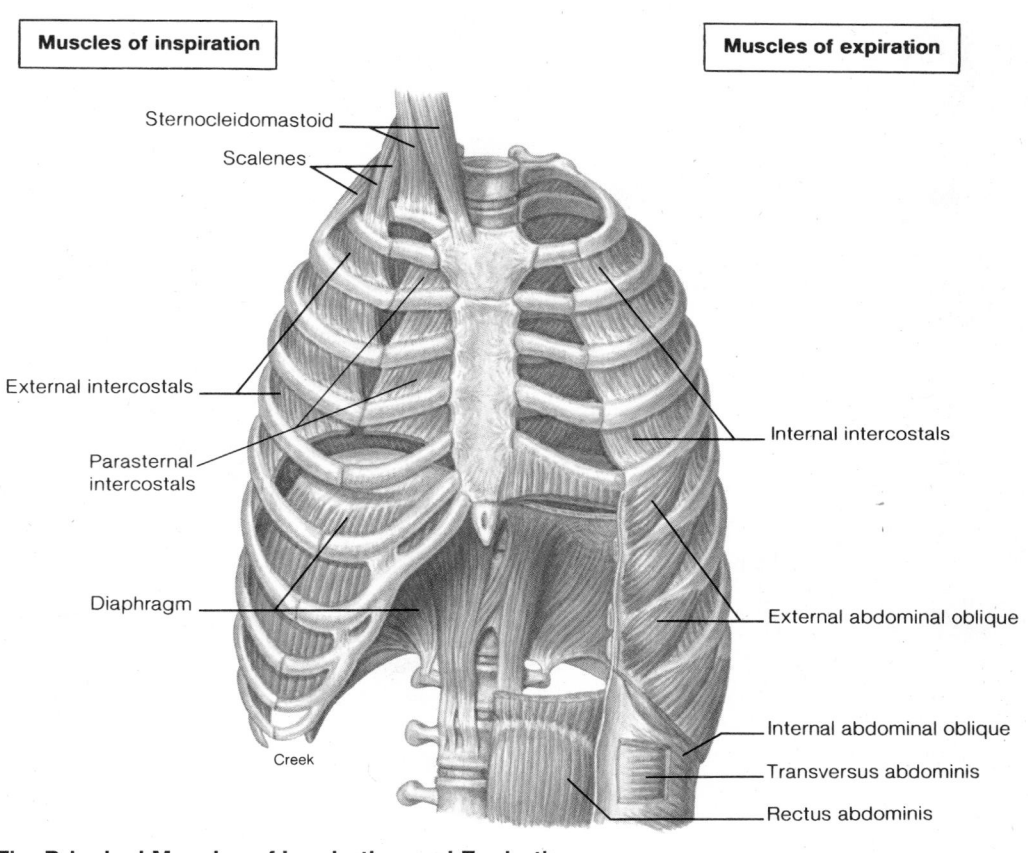

Muscles of inspiration

Sternocleidomastoid

Scalenes

External intercostals

Parasternal intercostals

Diaphragm

Creek

Muscles of expiration

Internal intercostals

External abdominal oblique

Internal abdominal oblique

Transversus abdominis

Rectus abdominis

The Principal Muscles of Inspiration and Expiration
Figure 16.14

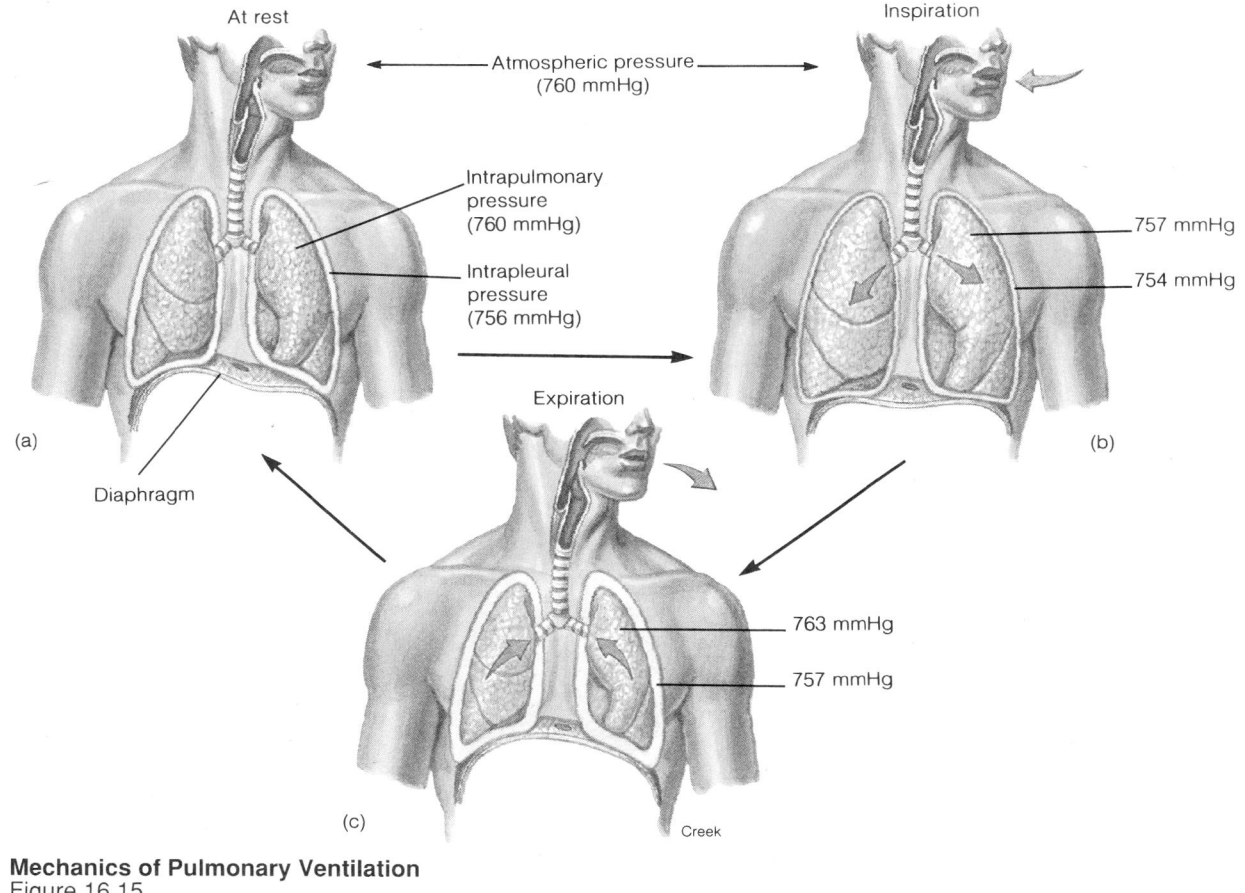

At rest

Atmospheric pressure
(760 mmHg)

Intrapulmonary
pressure
(760 mmHg)

Intrapleural
pressure
(756 mmHg)

(a)

Diaphragm

Inspiration

757 mmHg

754 mmHg

(b)

Expiration

763 mmHg

757 mmHg

(c)

Creek

Mechanics of Pulmonary Ventilation
Figure 16.15

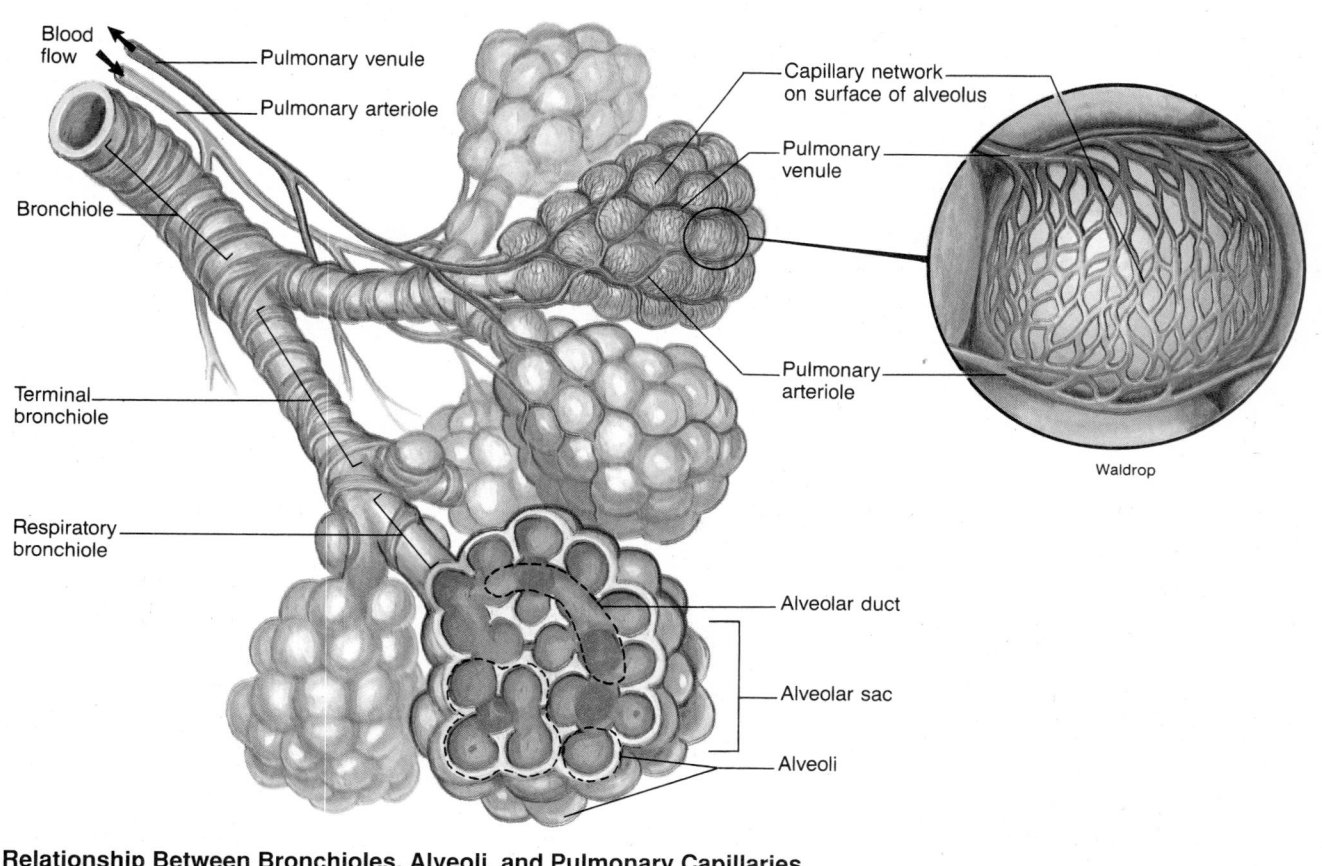

Blood flow

Pulmonary venule

Pulmonary arteriole

Bronchiole

Terminal bronchiole

Respiratory bronchiole

Capillary network on surface of alveolus

Pulmonary venule

Pulmonary arteriole

Waldrop

Alveolar duct

Alveolar sac

Alveoli

Relationship Between Bronchioles, Alveoli, and Pulmonary Capillaries
Figure 16.22

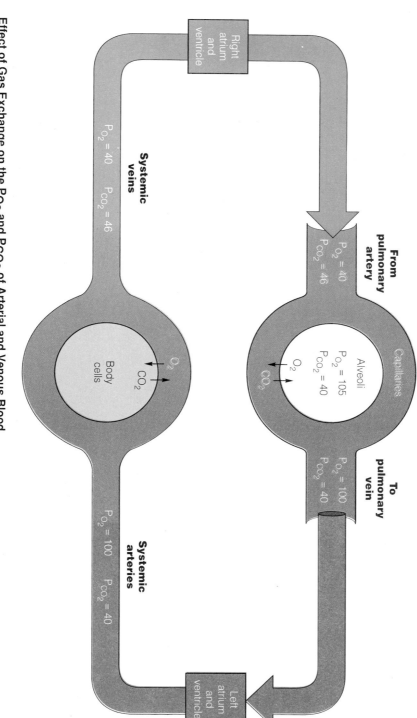

Effect of Gas Exchange on the PO_2 and PCO_2 of Arterial and Venous Blood
Figure 16.24

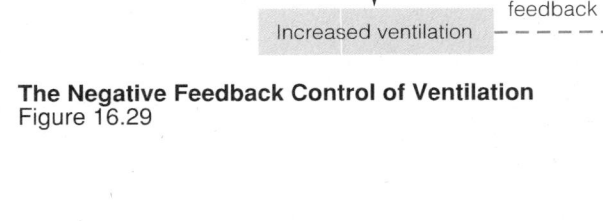

Decreased ventilation

Increased arterial P_{CO_2} ⊖

Plasma CO_2

↓ Blood pH

Blood

CSF

Chemoreceptors
in medulla oblongata

Chemoreceptors
in aortic and
carotid bodies

Respiratory center
in medulla oblongata

Sensory
neurons

Spinal cord
motor neurons

**Respiratory
muscles**

Negative
feedback

Increased ventilation

The Negative Feedback Control of Ventilation
Figure 16.29

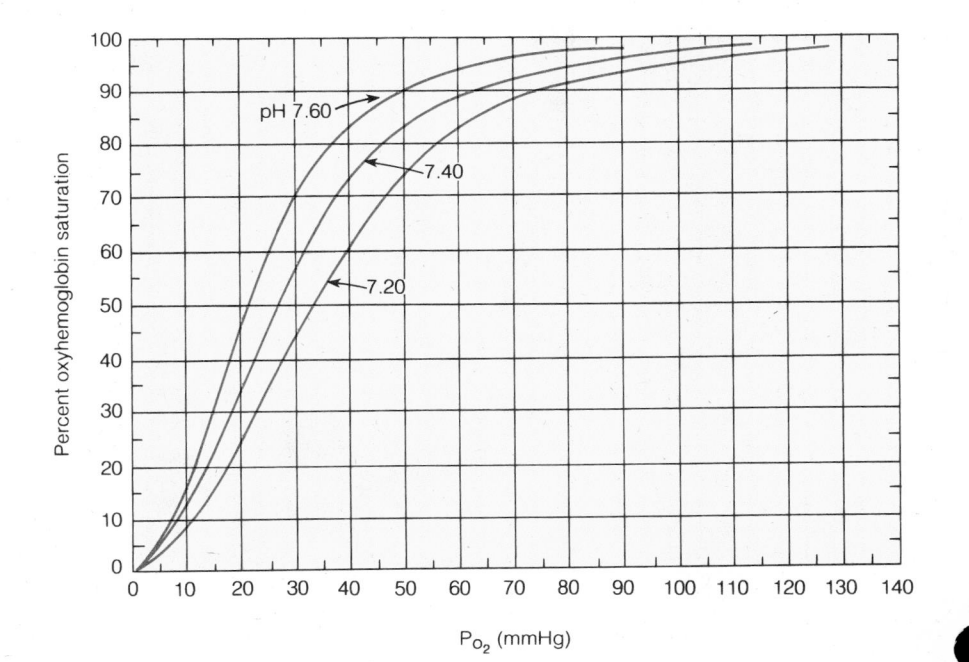

Effect of pH on the Oxyhemoglobin Dissociation Curve
Figure 16.34

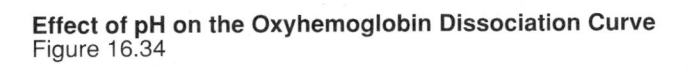

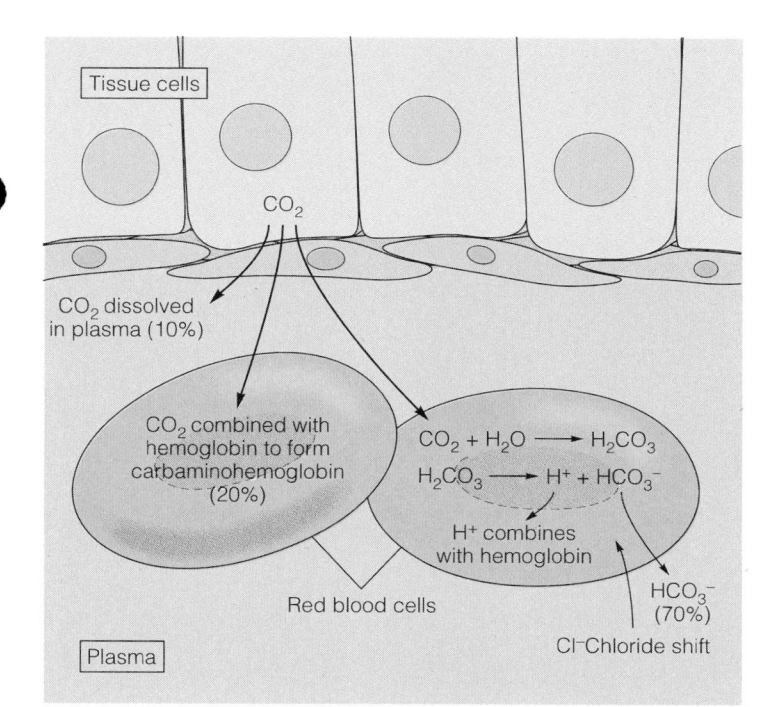

Carbon Dioxide Transport and the Chloride Shift in Tissue Capillaries
Figure 16.37

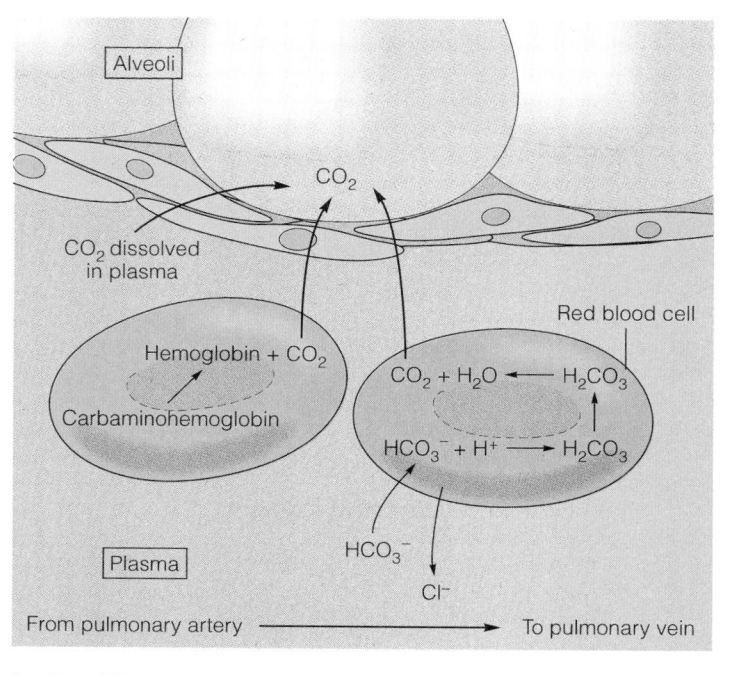

Carbon Dioxide Transport in Pulmonury Capillaries
Figure 16.38

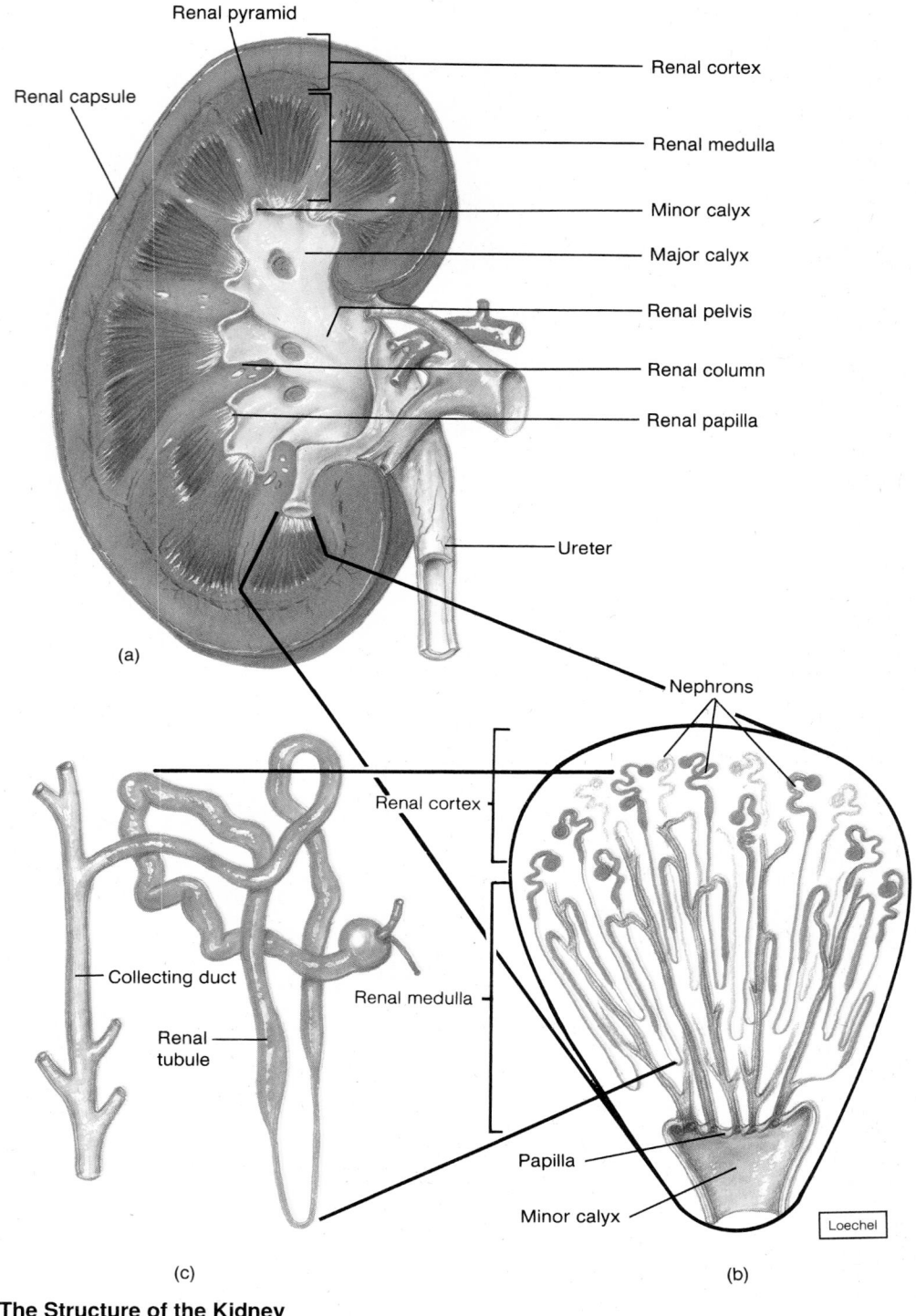

Renal pyramid

Renal cortex

Renal capsule

Renal medulla

Minor calyx

Major calyx

Renal pelvis

Renal column

Renal papilla

Ureter

(a)

Nephrons

Renal cortex

Collecting duct

Renal medulla

Renal tubule

Papilla

Minor calyx

Loechel

(c)

(b)

The Structure of the Kidney
Figure 17.2

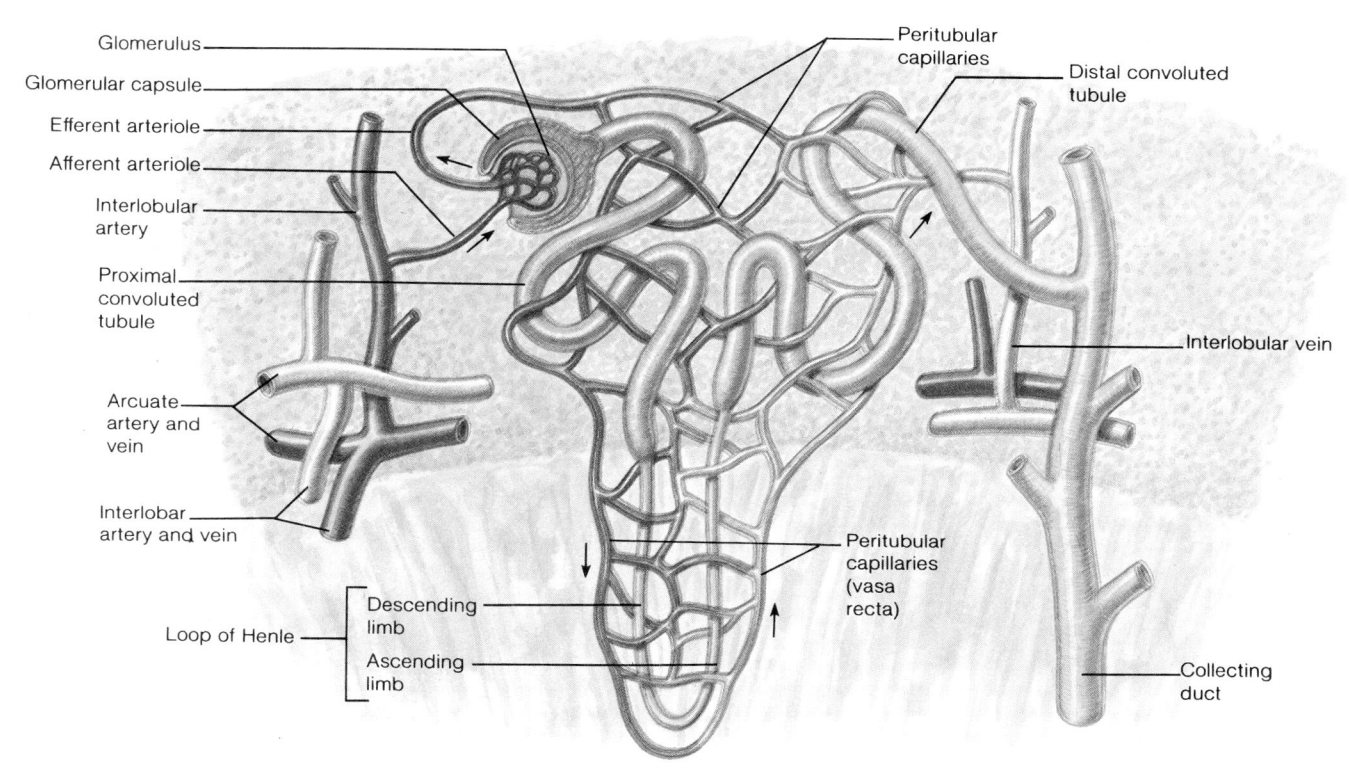

Glomerulus

Glomerular capsule

Efferent arteriole

Afferent arteriole

Interlobular artery

Proximal convoluted tubule

Arcuate artery and vein

Interlobar artery and vein

Loop of Henle

Descending limb

Ascending limb

Peritubular capillaries

Distal convoluted tubule

Interlobular vein

Peritubular capillaries (vasa recta)

Collecting duct

The Nephron Tubule and Its Associated Blood Vessels
Figure 17.5

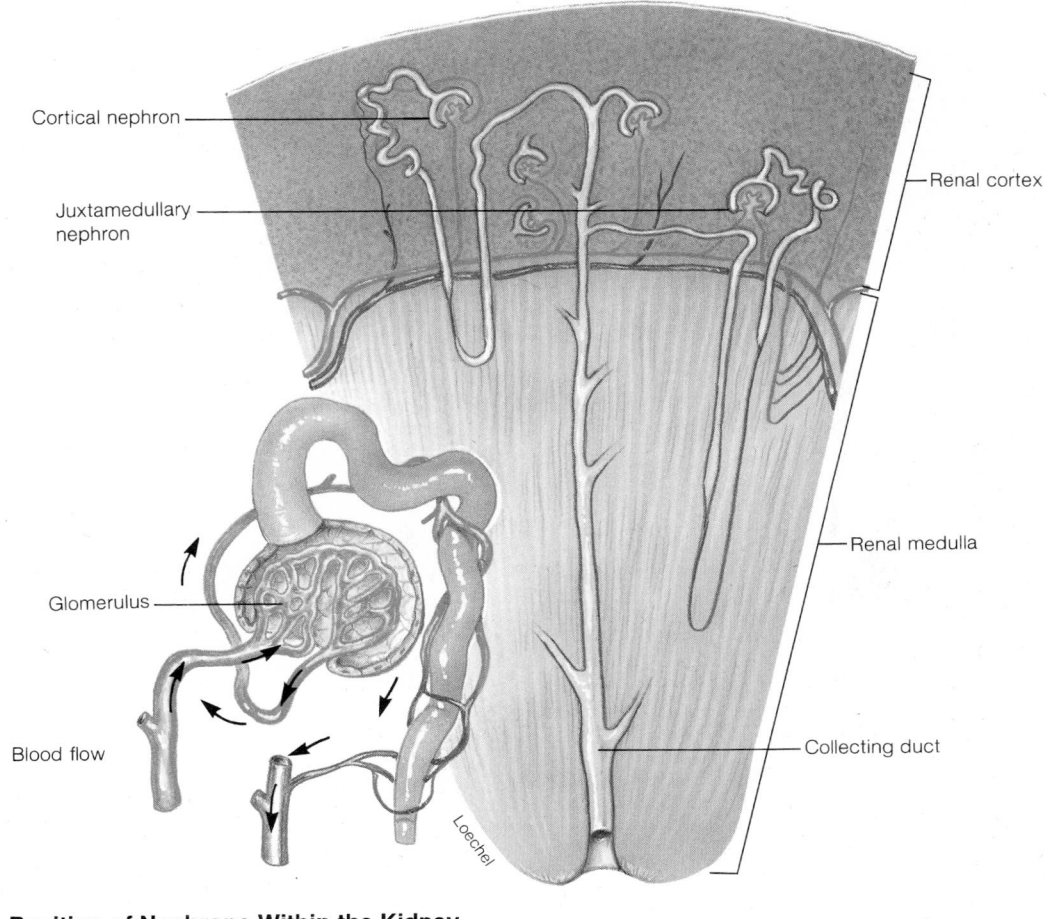

Cortical nephron

Juxtamedullary nephron

Glomerulus

Blood flow

Loechel

Renal cortex

Renal medulla

Collecting duct

Position of Nephrons Within the Kidney
Figure 17.6

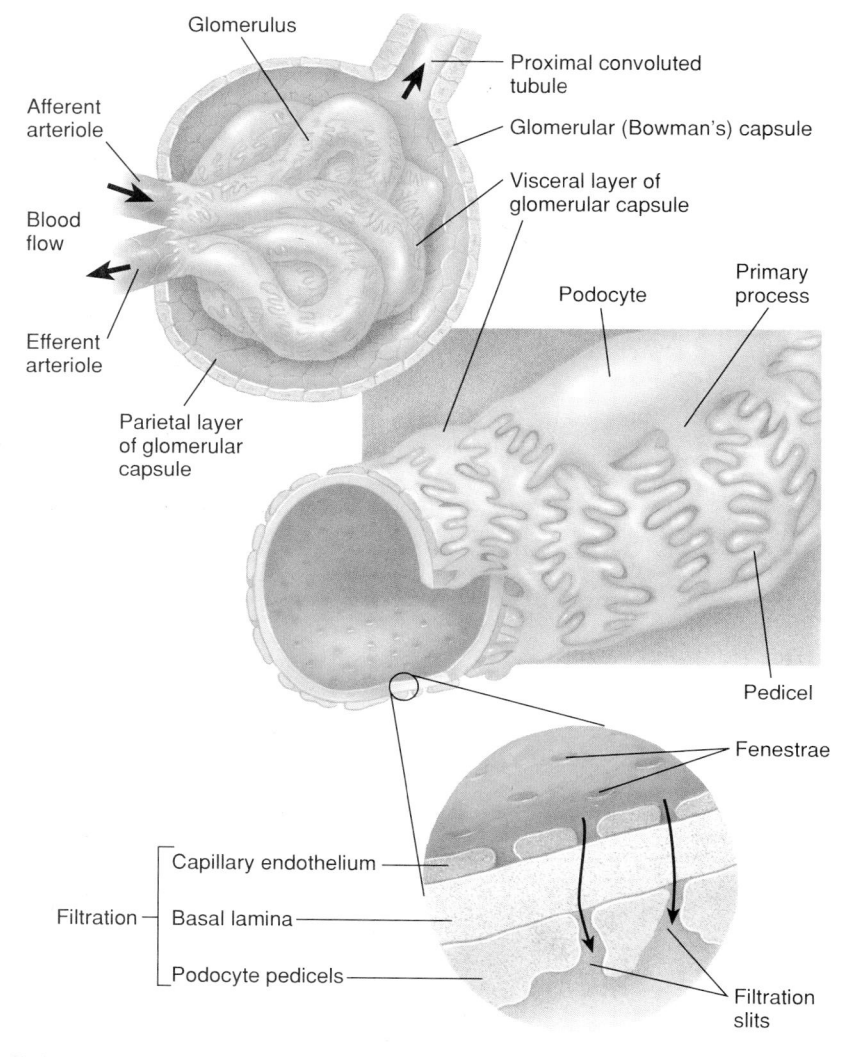

Glomerulus

Afferent arteriole

Blood flow

Efferent arteriole

Proximal convoluted tubule

Glomerular (Bowman's) capsule

Visceral layer of glomerular capsule

Parietal layer of glomerular capsule

Podocyte

Primary process

Pedicel

Fenestrae

Capillary endothelium

Filtration — Basal lamina

Podocyte pedicels

Filtration slits

Relationship Between Glomerular Capillaries and the Inner Layer of Bowman's Capsule
Figure 17.8

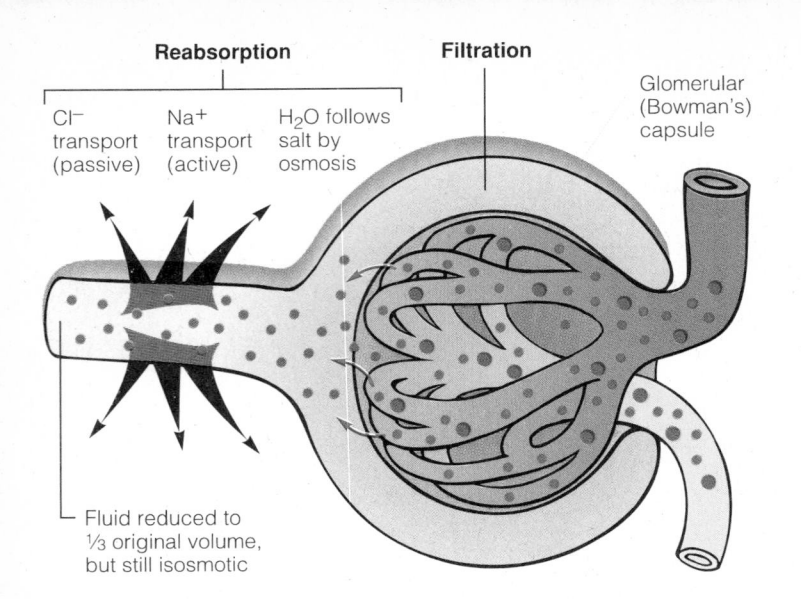

Reabsorption

Cl⁻ transport (passive) Na⁺ transport (active) H₂O follows salt by osmosis

Filtration

Glomerular (Bowman's) capsule

Fluid reduced to ⅓ original volume, but still isosmotic

Mechanism of Salt and Water Reabsorption in the Proximal Tubule
Figure 17.14

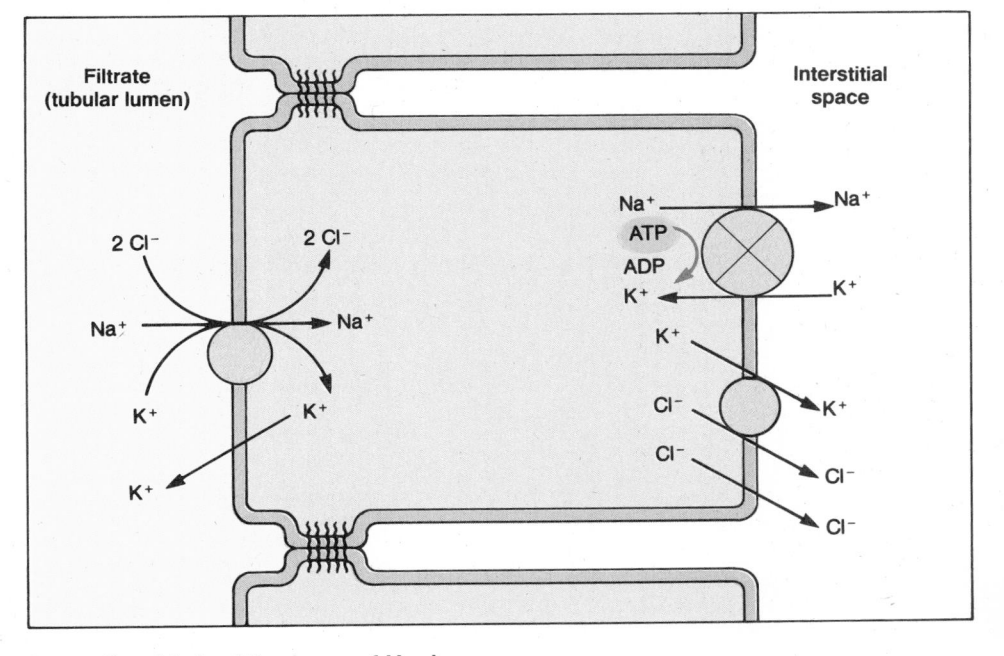

Filtrate (tubular lumen)

Interstitial space

2 Cl⁻   2 Cl⁻

Na⁺   Na⁺

ATP

ADP

K⁺   K⁺

K⁺   K⁺

K⁺

K⁺   K⁺

Cl⁻

Cl⁻   Cl⁻

Cl⁻

Ascending Limb of the Loop of Henle
Figure 17.15

(a)

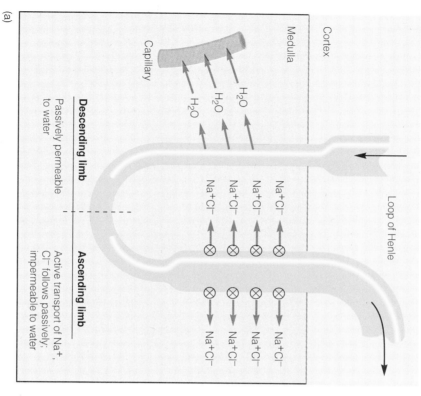

(b)

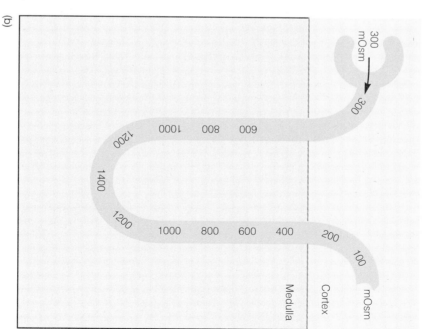

The Countercurrent Multiplier System
Figure 17.16

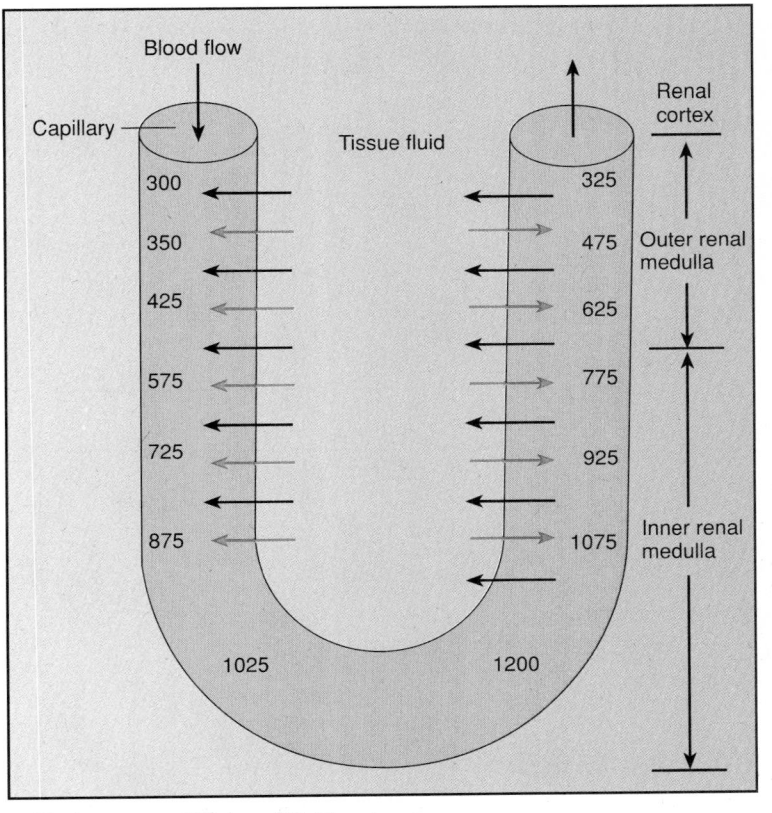

Black arrows = diffusion of NaCl and urea
Blue arrows = movement of water by osmosis

Countercurrent Exchange in the Vasa Recta
Figure 17.17

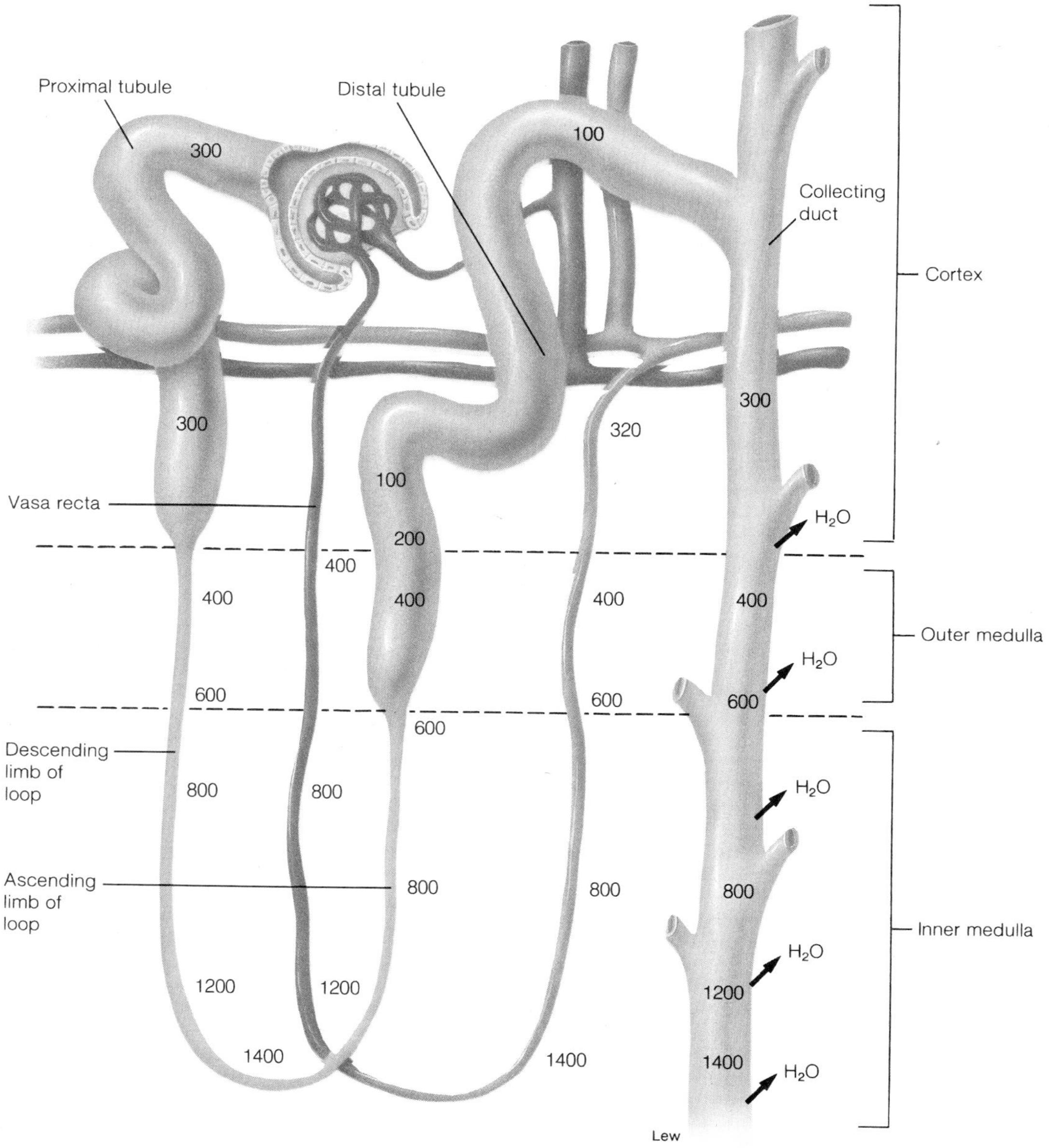

Osmolality of Different Regions of the Nephron Tubules
Figure 17.19

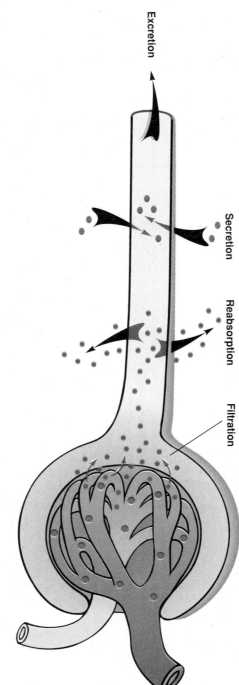

Excretion

Secretion

Reabsorption

Filtration

Filtration, Reabsorption, and Secretion
Figure 17.20

1

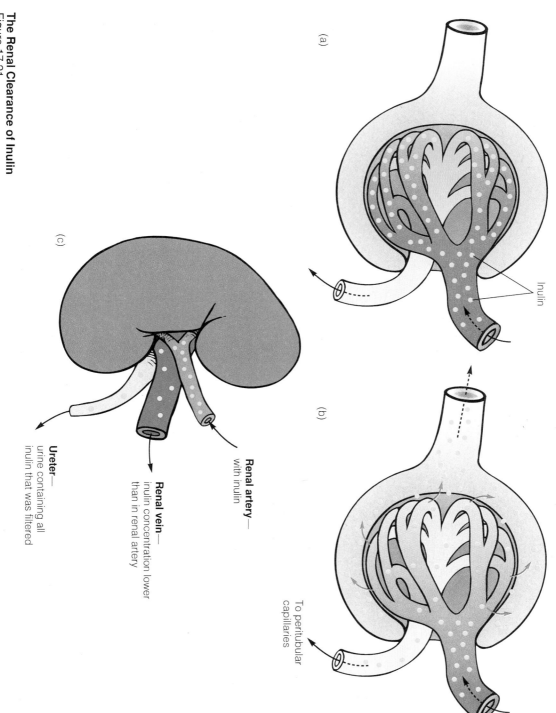

(a)

(b)

(c)

Inulin

To peritubular
capillaries

Renal artery—
with inulin

Renal vein—
inulin concentration lower
than in renal artery

Ureter—
urine containing all
inulin that was filtered

The Renal Clearance of Inulin
Figure 17.21

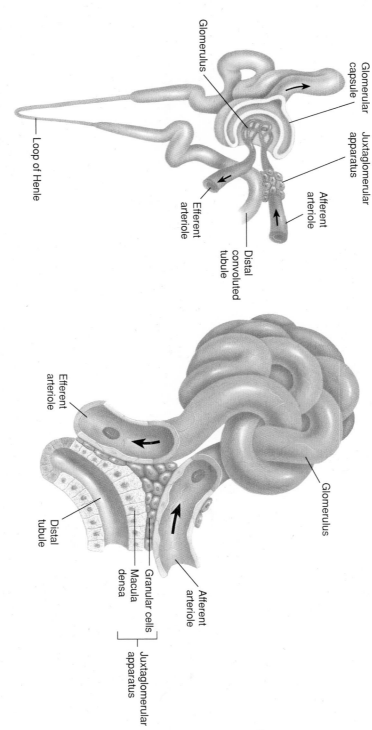
Glomerular capsule
Juxtaglomerular apparatus
Afferent arteriole
Distal convoluted tubule
Glomerulus
Efferent arteriole
Loop of Henle
Glomerulus
Afferent arteriole
Granular cells
Macula densa
Juxtaglomerular apparatus
Efferent arteriole
Distal tubule

The Juxtaglomerular Apparatus
Figure 17.24

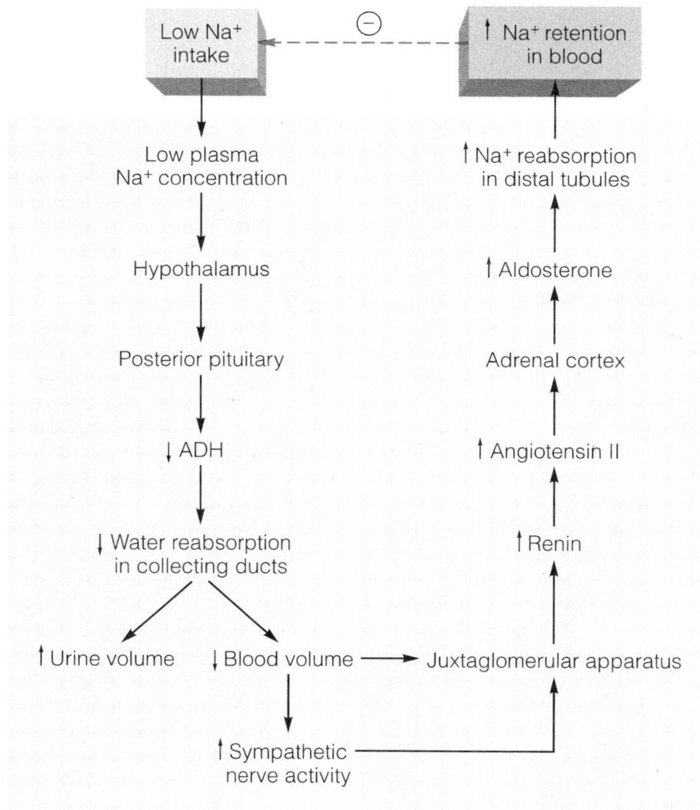

Low Na+ intake ⊖ ↑ Na+ retention in blood

Low Na+ intake → Low plasma Na+ concentration → Hypothalamus → Posterior pituitary → ↓ ADH → ↓ Water reabsorption in collecting ducts

↓ Water reabsorption in collecting ducts → ↑ Urine volume
↓ Water reabsorption in collecting ducts → ↓ Blood volume → Juxtaglomerular apparatus
↓ Blood volume → ↑ Sympathetic nerve activity → Juxtaglomerular apparatus

Juxtaglomerular apparatus → ↑ Renin → ↑ Angiotensin II → Adrenal cortex → ↑ Aldosterone → ↑ Na+ reabsorption in distal tubules → ↑ Na+ retention in blood

Renal Compensations for Low Sodium Intake
Figure 17.25

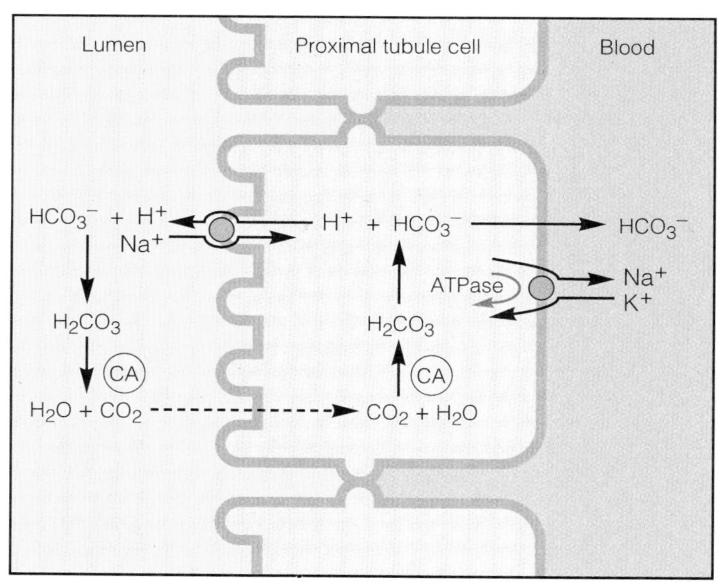

Bicarbonate Reabsorption by the Renal Nephron
Figure 17.28

107

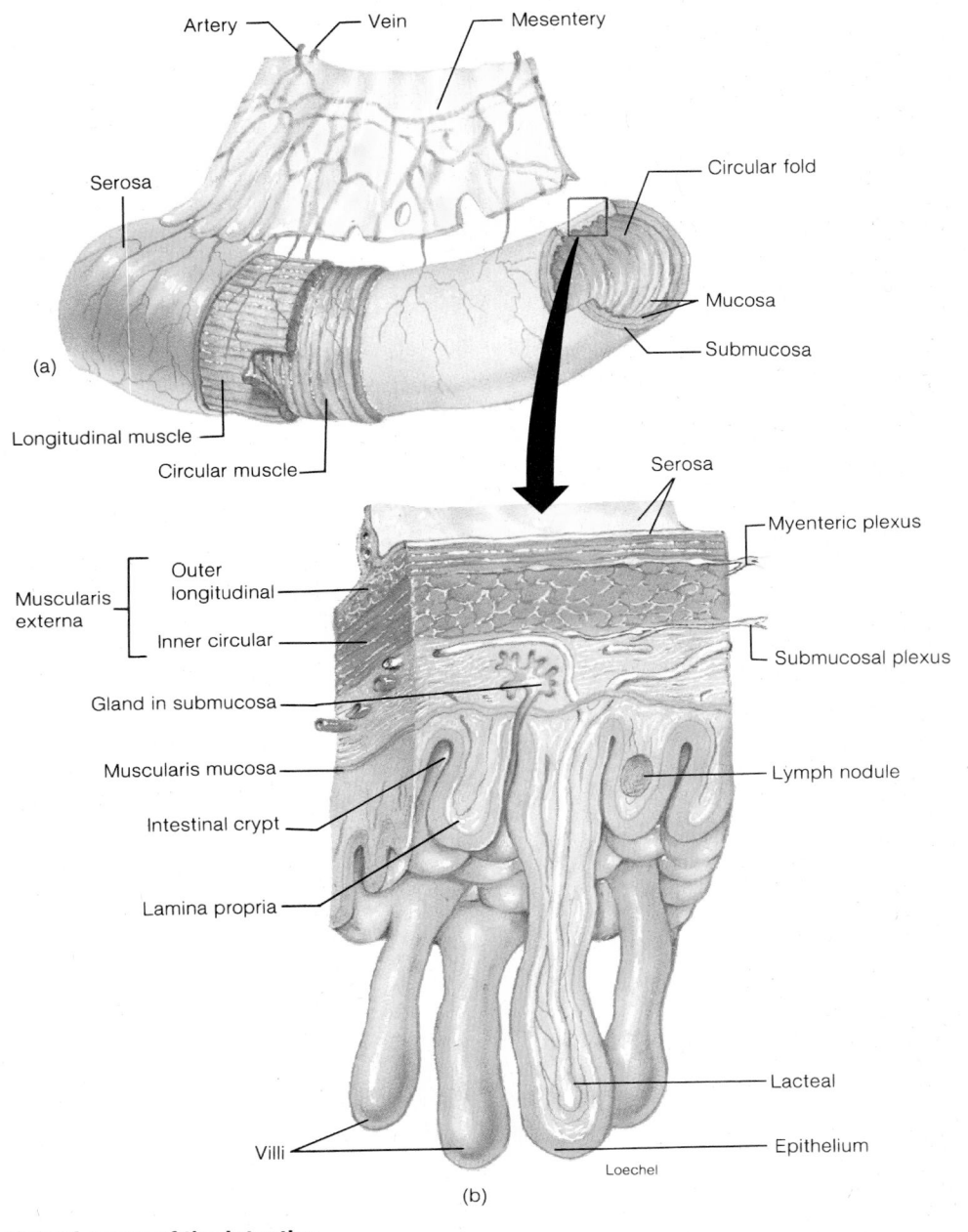
Artery
Vein
Mesentery
Serosa
Circular fold
Mucosa
Submucosa
(a)
Longitudinal muscle
Circular muscle
Serosa
Myenteric plexus
Muscularis externa
Outer longitudinal
Inner circular
Submucosal plexus
Gland in submucosa
Muscularis mucosa
Lymph nodule
Intestinal crypt
Lamina propria
Lacteal
Villi
Epithelium
Loechel
(b)

Major Layers of the Intestine
Figure 18.3

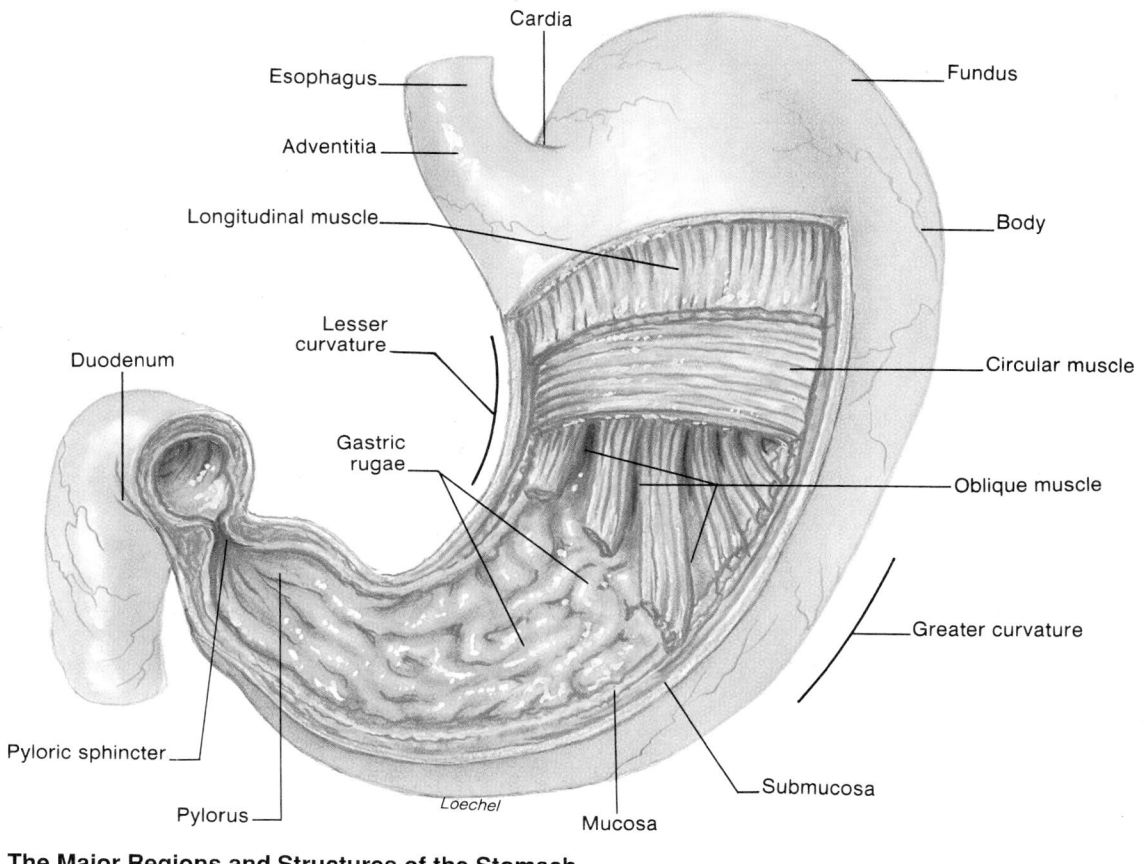

Cardia

Esophagus

Adventitia

Longitudinal muscle

Lesser curvature

Duodenum

Gastric rugae

Pyloric sphincter

Pylorus

Loechel

Mucosa

Fundus

Body

Circular muscle

Oblique muscle

Greater curvature

Submucosa

The Major Regions and Structures of the Stomach
Figure 18.5

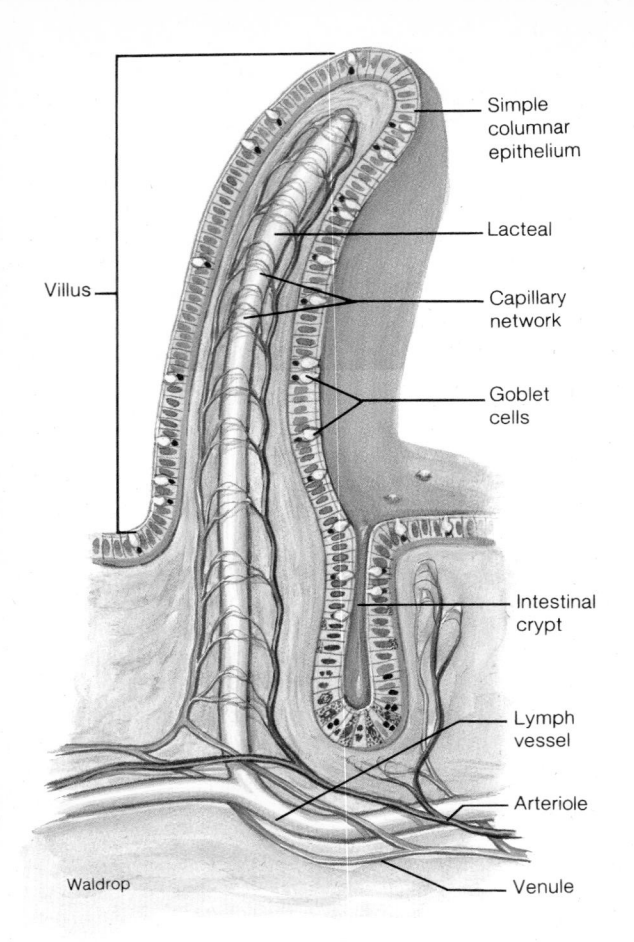

Simple
columnar
epithelium

Lacteal

Capillary
network

Goblet
cells

Villus

Intestinal
crypt

Lymph
vessel

Arteriole

Venule

Waldrop

Structure of an Intestinal Villus
Figure 18.13

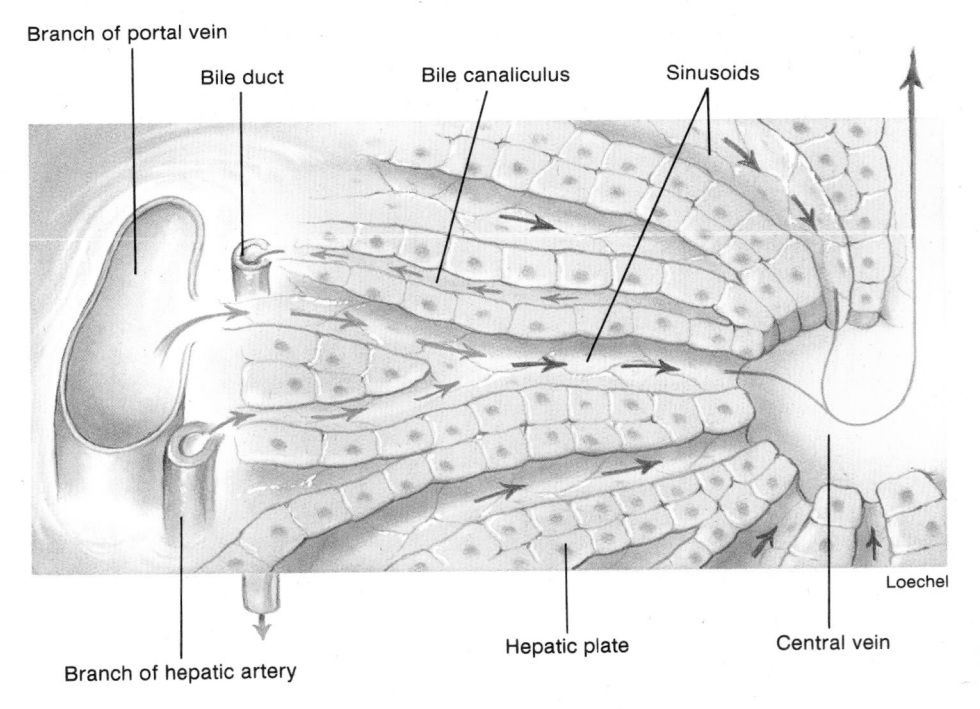

Branch of portal vein

Bile duct

Bile canaliculus

Sinusoids

Hepatic plate

Central vein

Loechel

Branch of hepatic artery

Flow of Blood and Bile in a Liver Lobule
Figure 18.21

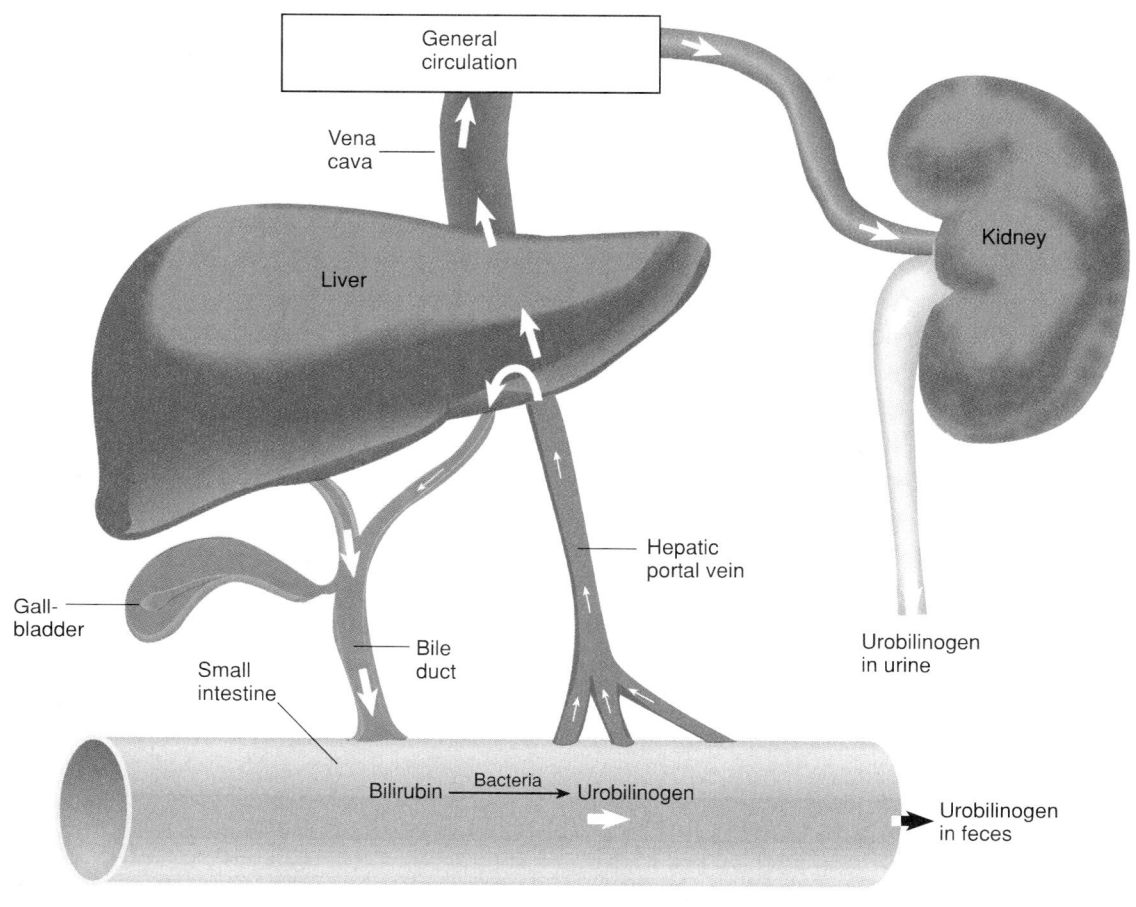

The Enterohepatic Circulation of Urobilinogen
Figure 18.23

From liver and gallbladder

Bile duct

Micelles of bile salts, cholesterol, and lecithin

1

From stomach

2   Free fatty acids

+ Lipase

Monoglycerides

3

Into micelles

Fat droplets (triglycerides)

Emulsified fat droplets (triglycerides)

Step 1 Emulsification of fat droplets by bile salts

Step 2 Hydrolysis of triglycerides in emulsified fat droplets into fatty acid and monoglycerides

Step 3 Dissolving of fatty acids and monoglycerides into micelles to produce ''mixed micelles''

Steps in the Digestion of Fat
Figure 18.33

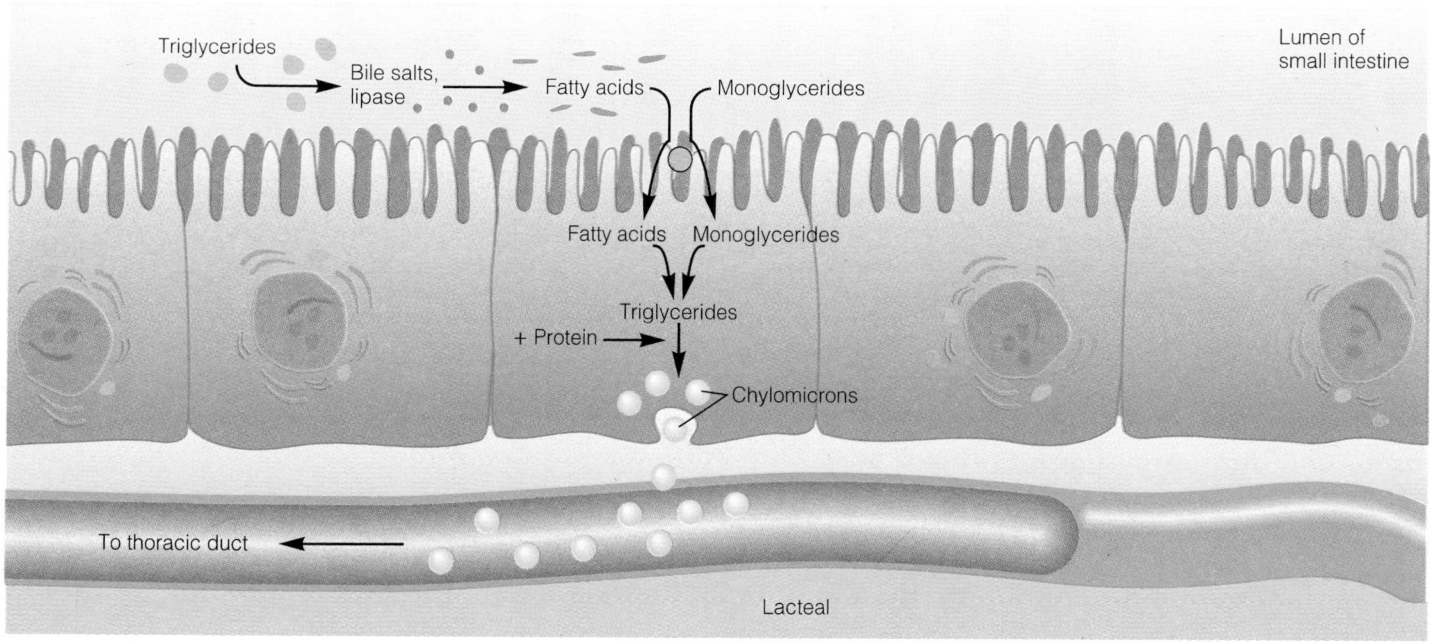

The Absorption of Fat
Figure 18.34

To thoracic duct

Lacteal

Triglycerides

Bile salts, lipase

Fatty acids

+ Protein

Triglycerides

Fatty acids

Monoglycerides

Monoglycerides

Chylomicrons

Lumen of small intestine

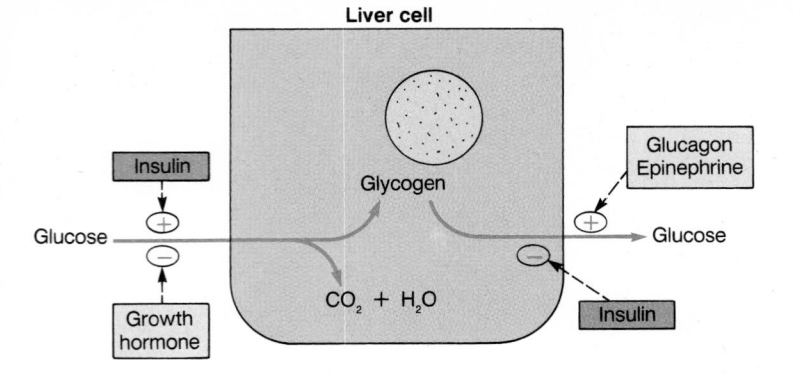

Liver cell

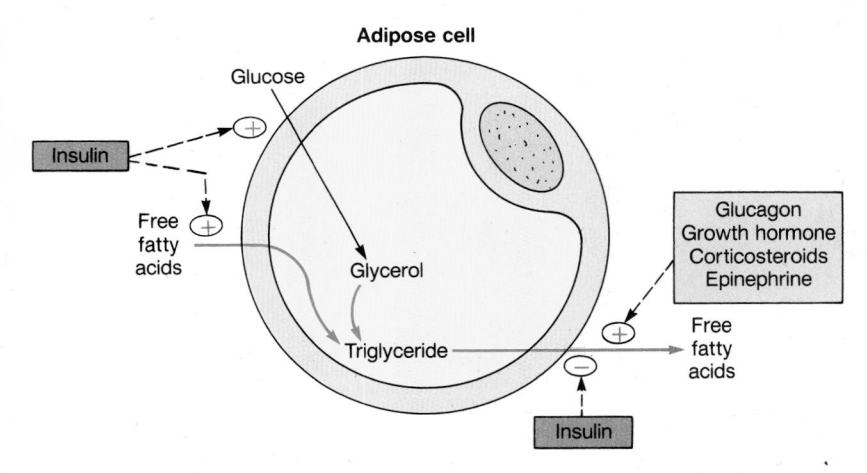

Adipose cell

Different Hormones Participate in the Regulation of Metabolism
Figure 19.3

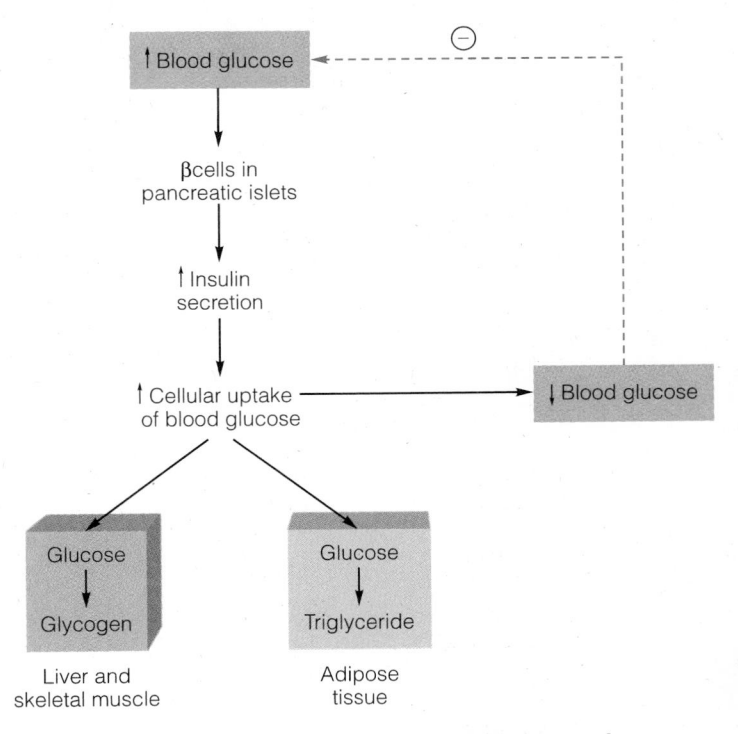

Insulin Regulation of the Blood Glucose Concentration
Figure 19.7

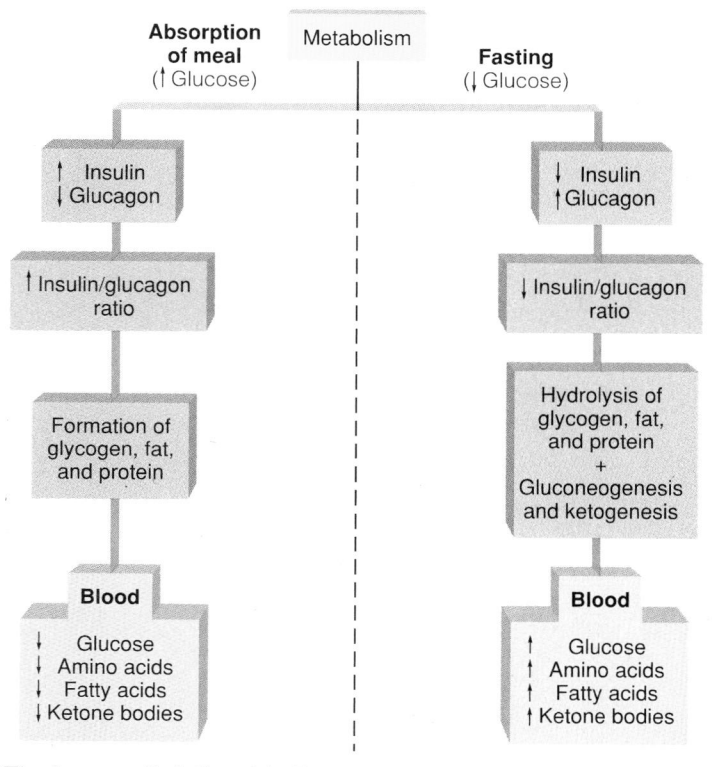

The Inverse Relationship Between Insulin and Glucagon Secretion and Action
Figure 19.9

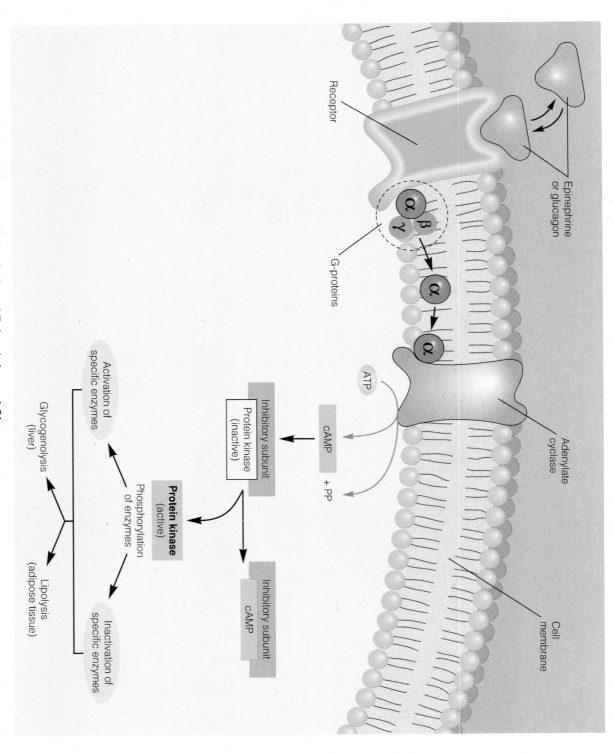

Cyclic AMP as a Second Messenger in the Action of Epinephrine and Glucagon
Figure 19.13

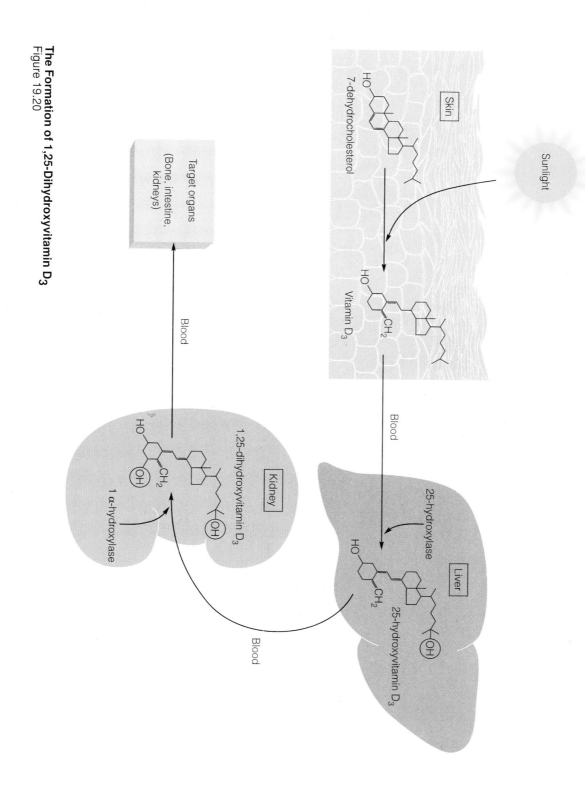

The Formation of 1,25-Dihydroxyvitamin D₃
Figure 19.20

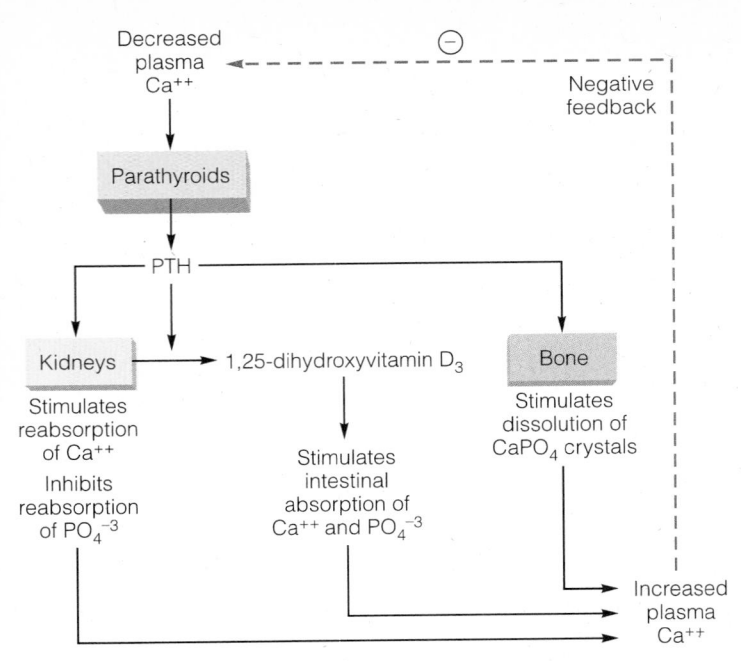

The Negative Feedback Regulation of the Blood Ca++ Concentration
Figure 19.22

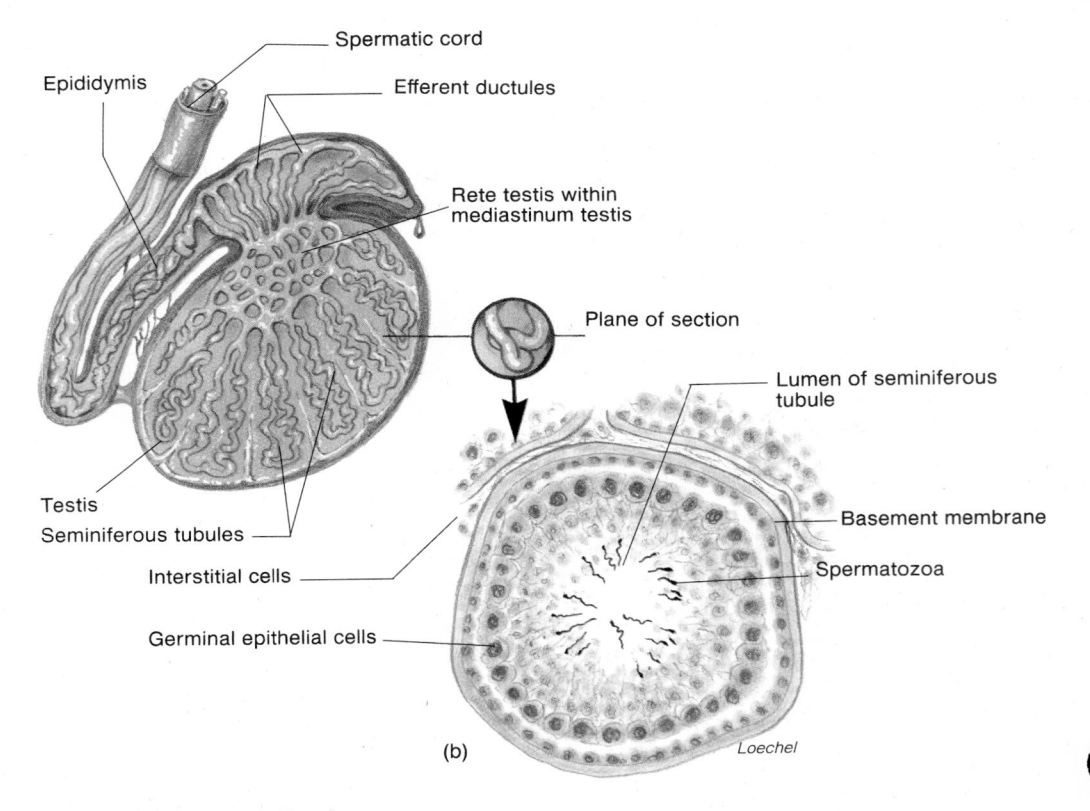

(b)

Loechel

The Structure of the Testis
Figure 20.12

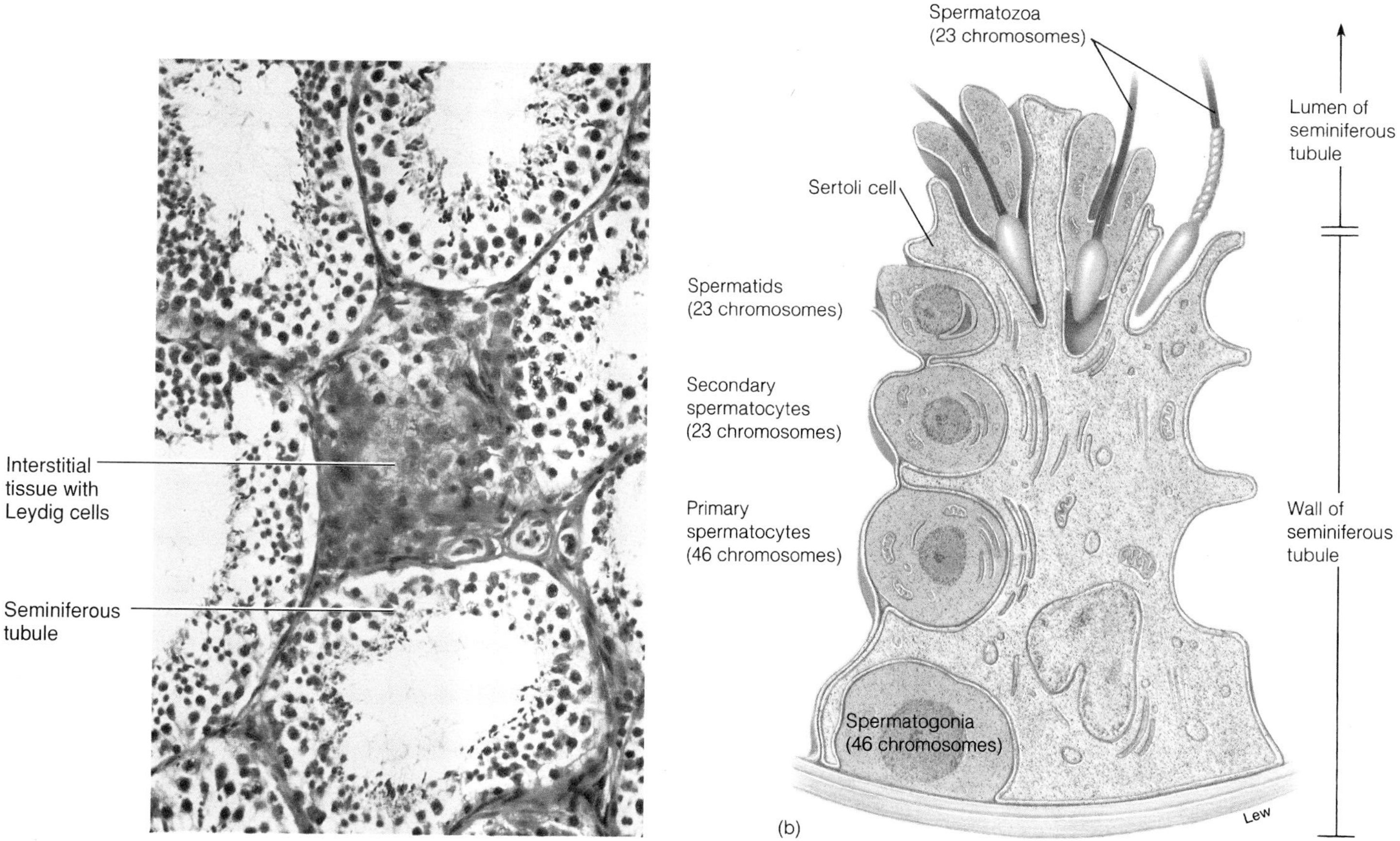
Interstitial tissue with Leydig cells
Seminiferous tubule
Spermatozoa (23 chromosomes)
Sertoli cell
Spermatids (23 chromosomes)
Secondary spermatocytes (23 chromosomes)
Primary spermatocytes (46 chromosomes)
Spermatogonia (46 chromosomes)
Lumen of seminiferous tubule
Wall of seminiferous tubule
(b)
Lew

Spermatogenesis Within the Semeniferous Tubules
Figure 20.17

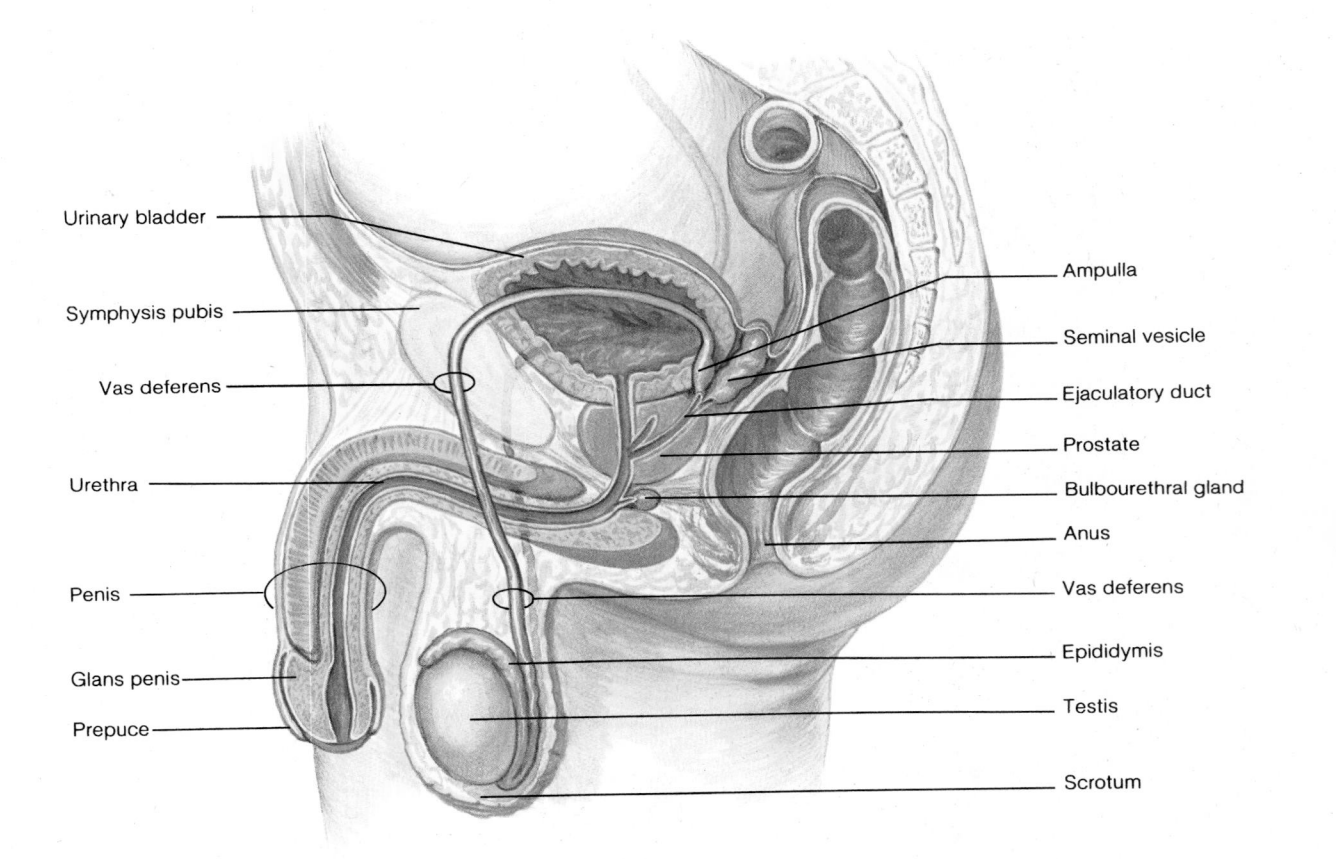

Urinary bladder

Symphysis pubis

Vas deferens

Urethra

Penis

Glans penis

Prepuce

Ampulla

Seminal vesicle

Ejaculatory duct

Prostate

Bulbourethral gland

Anus

Vas deferens

Epididymis

Testis

Scrotum

Organs of the Male Reproductive System
Figure 20.21

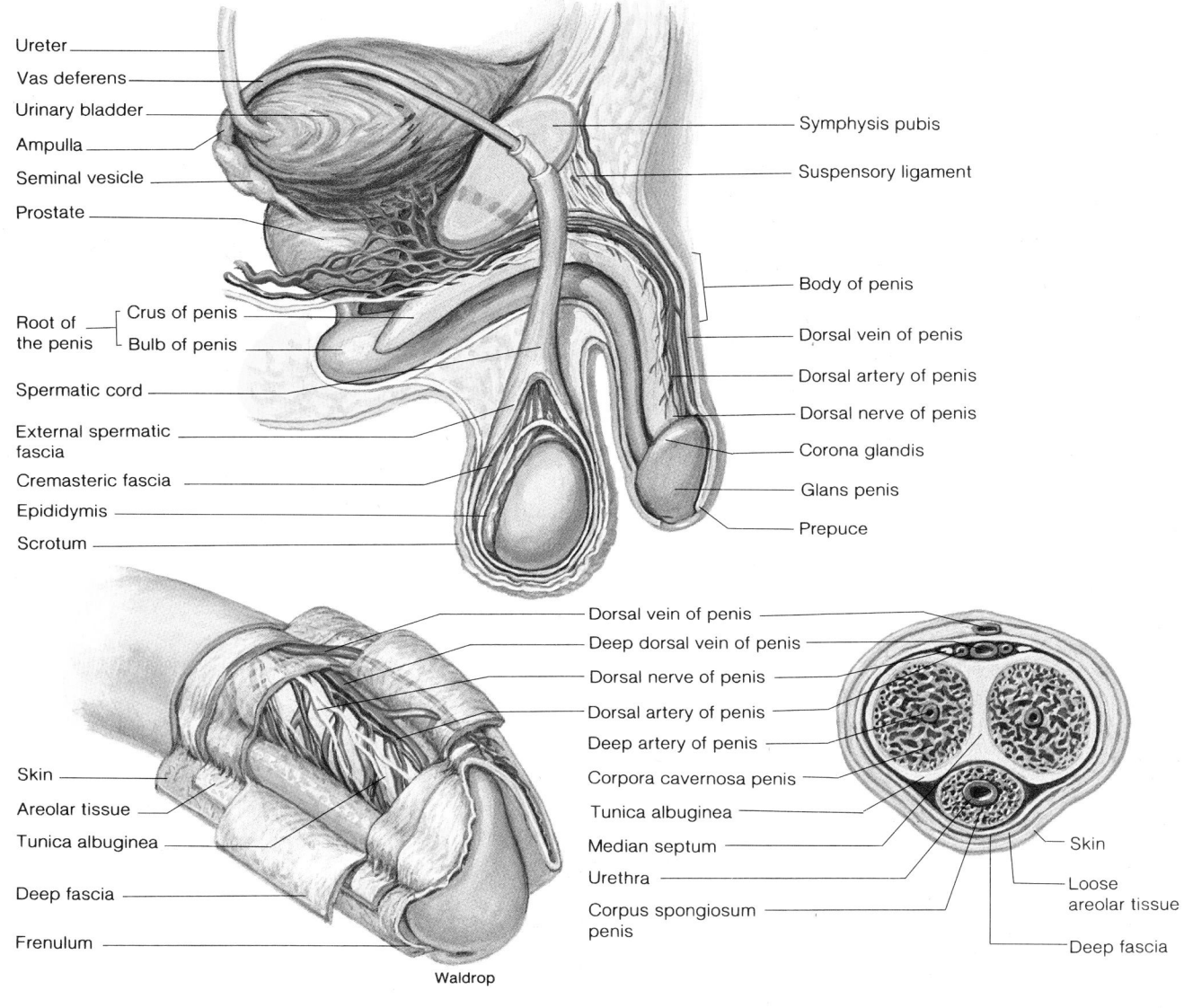

Ureter

Vas deferens

Urinary bladder

Ampulla

Seminal vesicle

Prostate

Root of the penis
Crus of penis
Bulb of penis

Spermatic cord

External spermatic fascia

Cremasteric fascia

Epididymis

Scrotum

Symphysis pubis

Suspensory ligament

Body of penis

Dorsal vein of penis

Dorsal artery of penis

Dorsal nerve of penis

Corona glandis

Glans penis

Prepuce

Skin

Areolar tissue

Tunica albuginea

Deep fascia

Frenulum

Dorsal vein of penis

Deep dorsal vein of penis

Dorsal nerve of penis

Dorsal artery of penis

Deep artery of penis

Corpora cavernosa penis

Tunica albuginea

Median septum

Urethra

Corpus spongiosum penis

Skin

Loose areolar tissue

Deep fascia

Waldrop

Structure of the Penis
Figure 20.22

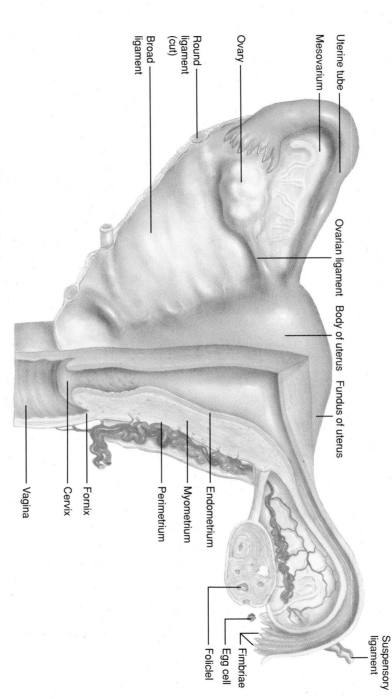

A Dorsal View of Female Reproductive Organs
Figure 20.24

Uterine tube

Mesovarium

Ovary

Round ligament (cut)

Broad ligament

Ovarian ligament

Body of uterus

Fundus of uterus

Suspensory ligament

Fimbriae

Egg cell

Folicel

Endometrium

Myometrium

Perimetrium

Fornix

Cervix

Vagina

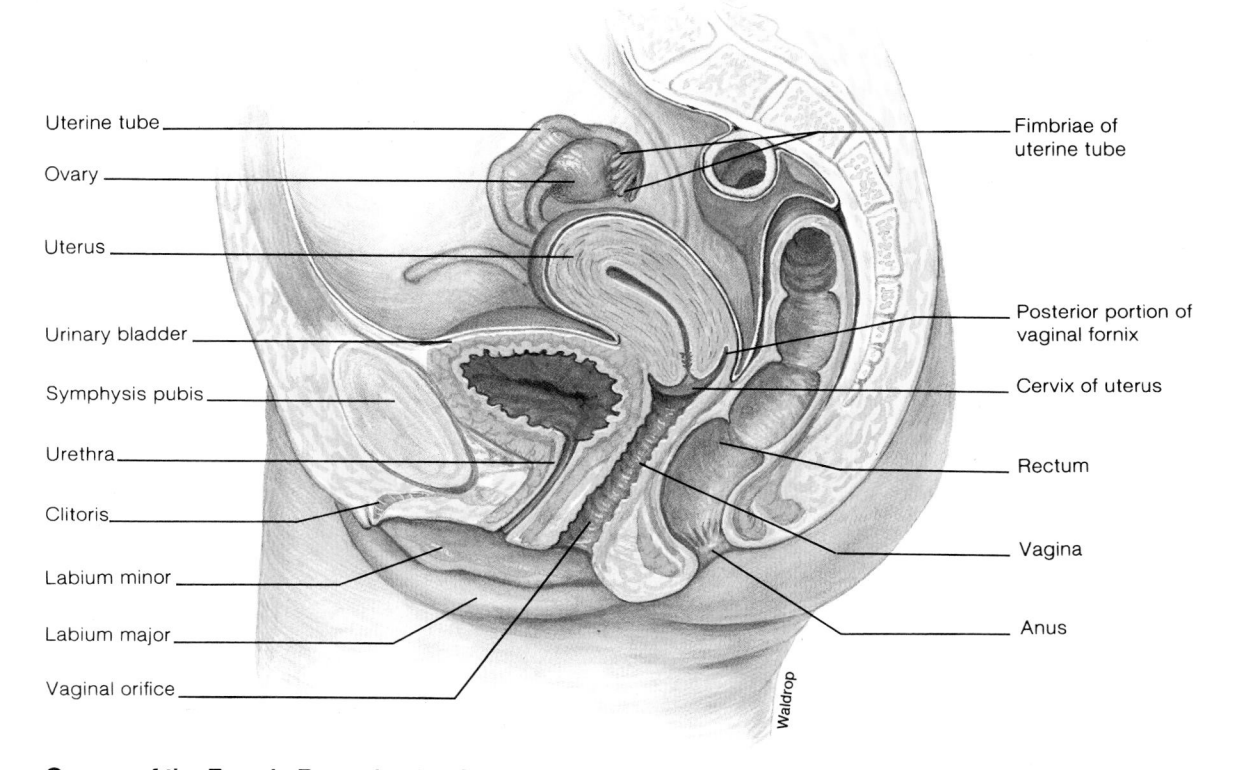

Uterine tube

Ovary

Uterus

Urinary bladder

Symphysis pubis

Urethra

Clitoris

Labium minor

Labium major

Vaginal orifice

Fimbriae of uterine tube

Posterior portion of vaginal fornix

Cervix of uterus

Rectum

Vagina

Anus

Waldrop

Organs of the Female Reproductive System
Figure 20.25

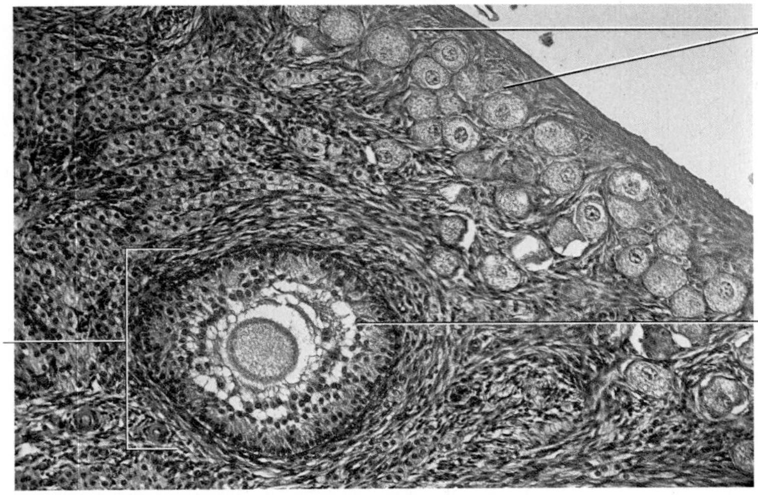

(a)

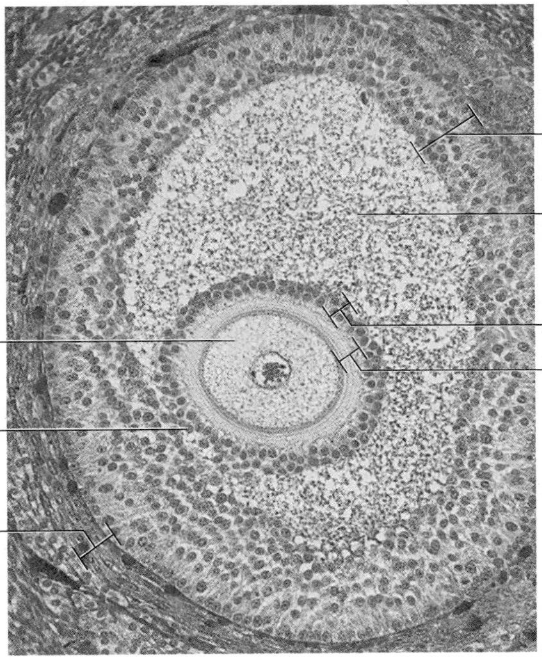

(b)

Photomicrographs of Primary and Graafian Follicles
Figure 20.27

124

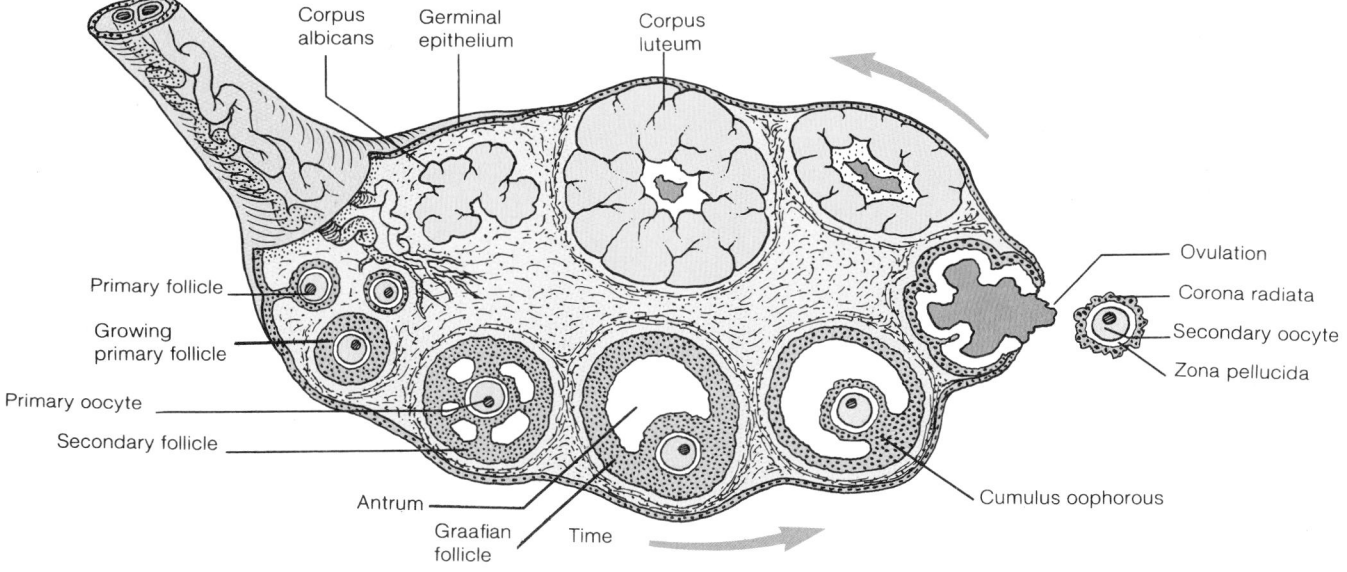

Diagram of the Cyclic Changes Within an Ovary
Figure 20.32

Corpus albicans

Germinal epithelium

Corpus luteum

Ovulation

Corona radiata

Secondary oocyte

Zona pellucida

Primary follicle

Growing primary follicle

Primary oocyte

Secondary follicle

Antrum

Graafian follicle

Time

Cumulus oophorous

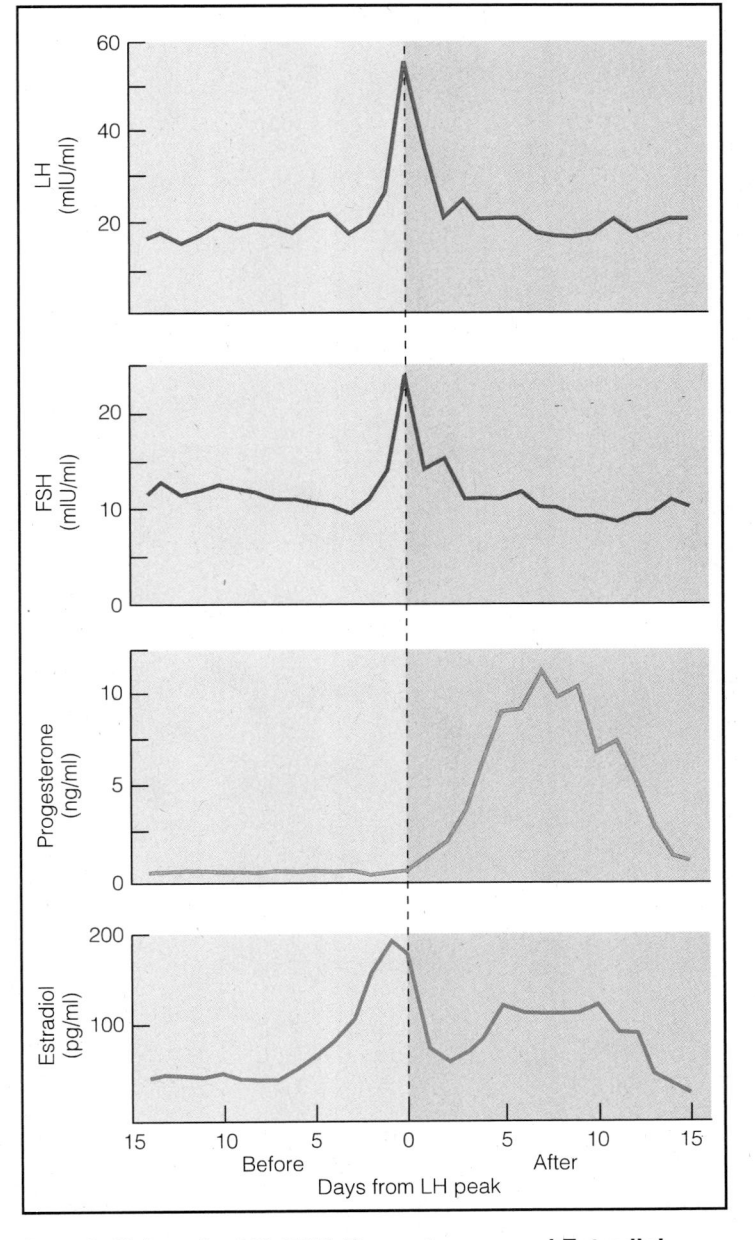

Sample Values for LH, FSH, Progesterone, and Estradiol During the Menstrual Cycle
Figure 20.33

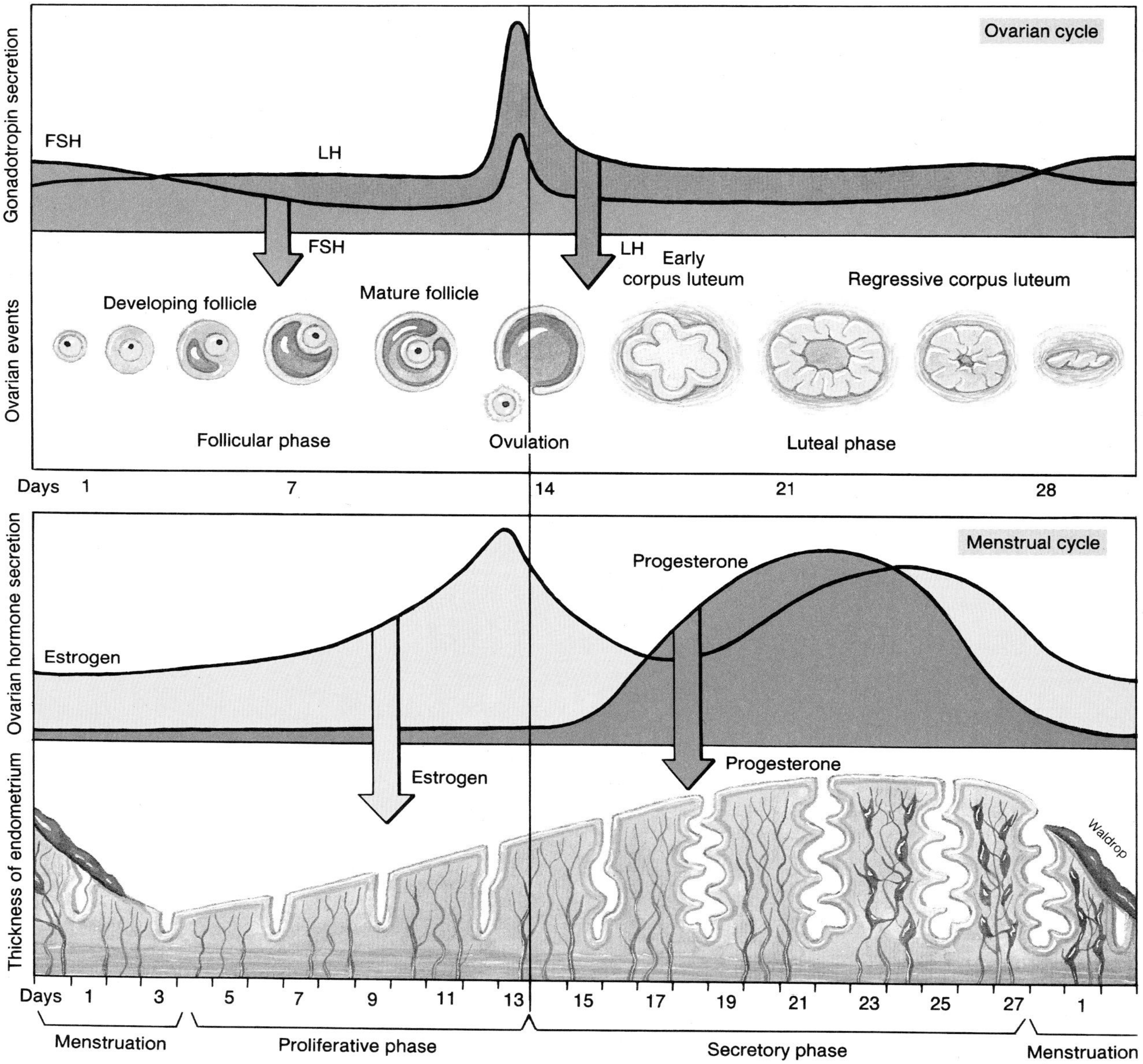

The Cycle of Ovulation and Menstruation
Figure 20.34

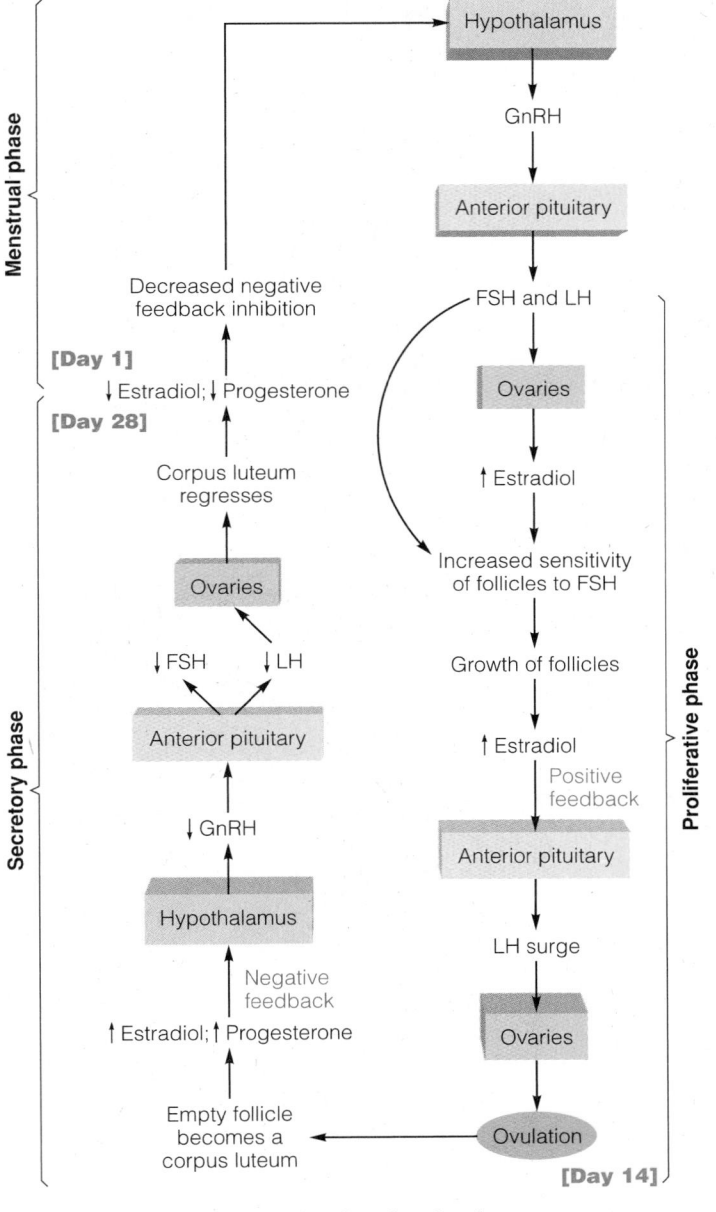

The Endocrine Control of the Ovarian Cycle
Figure 20.36

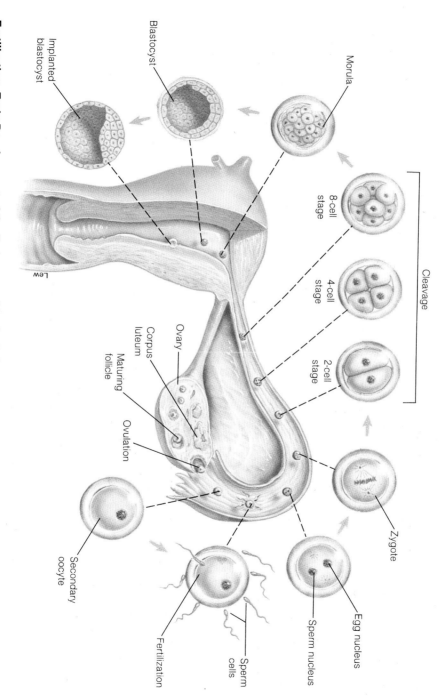

Fertilization, Early Development of the Embryo, and Implantation
Figure 20.42

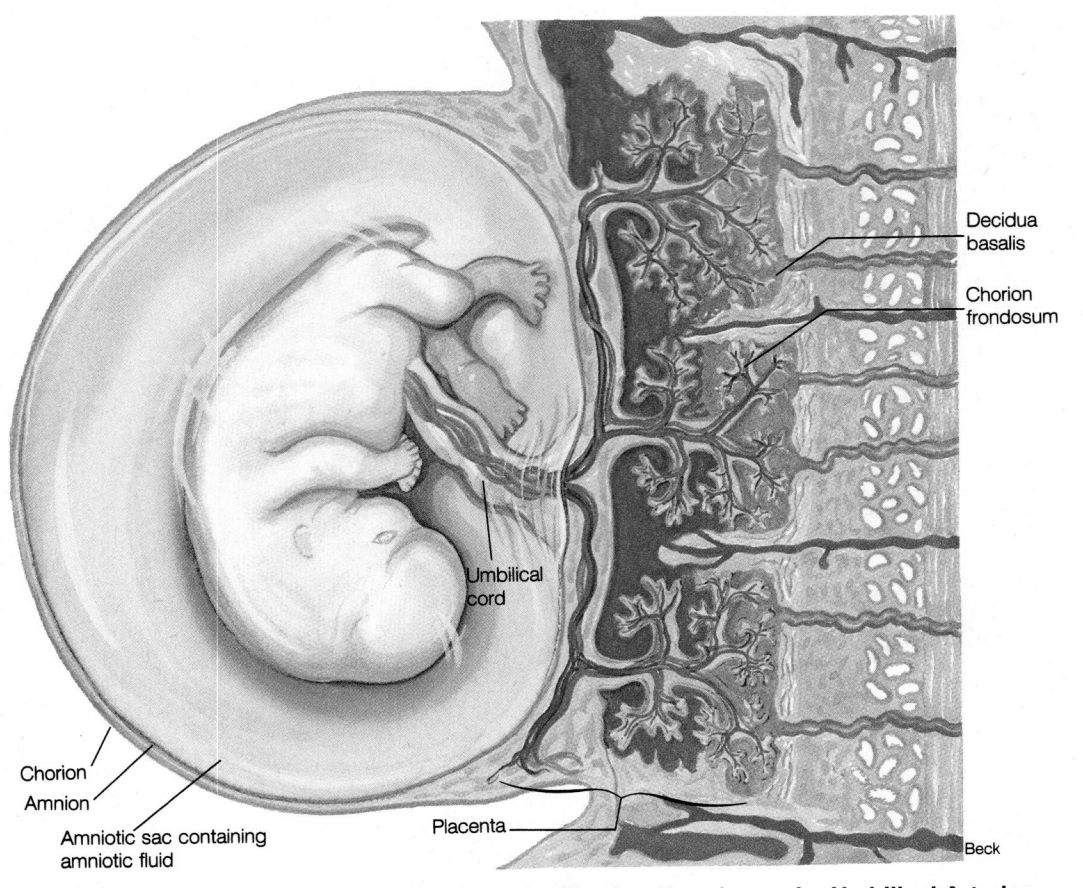

Blood from the Fetus is Carried to and from the Chorion Frondosum by Umbilical Arteries and Veins
Figure 20.47

Line Art

Fig. 10.18 From John W. Hole, Jr., *Human Anatomy and Physiology,* 6th ed. Copyright © 1993 Wm. C. Brown Communications, Dubuque, Iowa. Reprinted by permission of Times Mirror Higher Education Group, Inc., Dubuque, Iowa. All Rights Reserved.

Fig. 10.36 From Kent M. Van De Graaff, *Human Anatomy,* 4th ed. Copyright © 1995 Wm. C. Brown Communications, Inc., Dubuque, Iowa. Reprinted by permission of Times Mirror Higher Education Group, Inc., Dubuque, Iowa. All Rights Reserved.

Fig. 11.9 From Kent M. Van De Graaff, *Human Anatomy,* 4th ed. Copyright © 1995 Wm. C. Brown Communications, Inc., Dubuque, Iowa. Reprinted by permission of Times Mirror Higher Education Group, Inc., Dubuque, Iowa. All Rights Reserved.

Fig. 13.6 Adapted from A. Marchand, "Case of the Month, Circulating Anticoagulants: Chasing the Diagnosis" in Diagnostic Medicine, June 1983, page 14. Copyright © 1984 Medical Economics Company, Inc., Oradell, NJ. Used with permission.

Fig. 14.17 Adapted P. Astrand and K. Rodahl, *Textbook of Work Physiology,* 3d ed. Copyright © 1986 McGraw-Hill, Inc., New York, NY. Used by permission of the authors.

Fig. 16.15 From Kent M. Van De Graaff, *Human Anatomy,* 4th ed. Copyright © 1995 Wm. C. Brown Communications, Inc., Dubuque, Iowa. Reprinted by permission of Times Mirror Higher Education Group, Inc., Dubuque, Iowa. All Rights Reserved.

Fig. 17.2 From John W. Hole, Jr., *Human Anatomy and Physiology,* 6th ed. Copyright © 1993 Wm. C. Brown Communications, Dubuque, Iowa. Reprinted by permission of Times Mirror Higher Education Group, Inc., Dubuque, Iowa. All Rights Reserved.

Fig. 17.6 From John W. Hole, Jr., *Human Anatomy and Physiology,* 6th ed. Copyright © 1993 Wm. C. Brown Communications, Dubuque, Iowa. Reprinted by permission of Times Mirror Higher Education Group, Inc., Dubuque, Iowa. All Rights Reserved.

Fig. 20.21 From John W. Hole, Jr., *Human Anatomy and Physiology,* 6th ed. Copyright © 1993 Wm. C. Brown Communications, Dubuque, Iowa. Reprinted by permission of Times Mirror Higher Education Group, Inc., Dubuque, Iowa. All Rights Reserved.

Fig. 20.25 From John W. Hole, Jr., *Human Anatomy and Physiology,* 6th ed. Copyright © 1993 Wm. C. Brown Communications, Dubuque, Iowa. Reprinted by permission of Times Mirror Higher Education Group, Inc., Dubuque, Iowa. All Rights Reserved.

Fig. 20.32 From Kent M. Van De Graaff, *Human Anatomy,* 4th ed. Copyright © 1995 Wm. C. Brown Communications, Inc., Dubuque, Iowa. Reprinted by permission of Times Mirror Higher Education Group, Inc., Dubuque, Iowa. All Rights Reserved.

Fig. 20.34 From John W. Hole, Jr., *Human Anatomy and Physiology,* 6th ed. Copyright © 1993 Wm. C. Brown Communications, Dubuque, Iowa. Reprinted by permission of Times Mirror Higher Education Group, Inc., Dubuque, Iowa. All Rights Reserved.

Photo Credits

Fig. 7.6 H. Webster, from Hubbard, John, "The Vertebrate Peripheral Nervous System", © Plenum Press, 1974.

Fig. 7.19 Gilula, Reverend Steimbach, *Nature,* 355:262–265 © MacMillan Journals Limited. From Leland G. Johnson, *Biology,* 2d ed. Copyright © 1987 Wm. C. Brown Communications, Dubuque, Iowa. Reprinted by permission of Times Mirror Higher Education Group, Inc., Dubuque, Iowa. All Rights Reserved.

Fig. 8.15 Times Mirror Higher Education Group, Inc./Karl Rubin, photographer.

Fig. 10.14 Courtesy Dean E. Hillman.

Fig. 12.6 © John D. Cunningham/Visuals Unlimited.

Fig. 12.9 Dr. H. E. Huxley.

Fig. 12.10a–b © Dr. H. E. Huxley.

Fig. 12.10c From R. G. Kessel and R. H. Kardon: *Tissues and Organs: A Text-Atlas of Scanning Electron Microscopy,* W. H. Freeman and Company © 1979.

Fig. 12.11 © Dr. H. E. Huxley.

Fig. 15.18 From Alan S. Rosenthal, *New England Journal of Medicine* 303:1153, 1980.

Fig. 20.17 © Biophoto Associates/Photo Researchers, Inc.

Fig. 20.27a&b © Edwin A. Reschke.